Contents

KU-677-530

iii

Chemistry and the Needs of Society

A SYMPOSIUM HELD AT IMPERIAL COLLEGE, LONDON, ON 2nd—5th APRIL, 1974

Organised by The Industrial Division
of The Chemical Society, as part of
the Annual Chemical Congress, 1974

SPECIAL PUBLICATION NO. 26

THE CHEMICAL SOCIETY,
BURLINGTON HOUSE, LONDON W1V 0BN

ISBN 0 85186 218 7

Printed in Great Britain by Alden & Mowbray Ltd
at the Alden Press, Oxford

Resources

This island is almost made of coal and surrounded by fish. Only an organising genius could produce a shortage of coal and fish in Great Britain at the same time.

Aneurin Bevan

Carbon and Hydrogen Sources— the Supplier

by P. I. Walters; British Petroleum Company

I THINK it is very timely for us to be discussing sources and uses of carbon and hydrogen. Certain recent events in the oil world, particularly oil production restrictions and embargoes, together with various consumer government controls, have caused acute concern over adequacy of supply. The quadrupling of oil prices similarly leaves unresolved many questions affecting international trade and world financial stability. As suppliers, the international oil companies are responsible for meeting demands from a number of markets, including chemicals. In 1973, naphtha for chemicals in Europe represented about 6% of total oil consumption, or more importantly, over 30% of total gasoline. The chemical industry is a vital part of today's industrial system, and some oil companies have a special and particular interest, not just as suppliers, but as users in their own petro-chemical operations. I hope to indicate the nature of the problems with which we are faced, and possible future trends.

Carbon and hydrogen are essential for many needs in society today. They are used in substantial quantities both in materials and food as well as fuel, to provide the energy needed for industrial processes, mechanisation, transport, and heating.

Sources of carbon and hydrogen are very numerous and some, for example carbonate rock and water, are vast compared with the carbon and hydrogen content of present estimates for recoverable fossil fuels (oil, gas, oil shale, tar sands and coal), as illustrated in Figure 1.

However, if carbon and hydrogen found in non-fossil forms are to be used, a large energy input is required. Therefore these sources will be of no real significance until non-fossil energy is relatively abundant. This is not likely to be the case until somewhere near the end of this century. Hence I propose to concentrate on the fossil fuels as carbon and hydrogen sources which, for at least the next 10 to 15 years, will need to serve the joint

3

SOURCES OF CARBON AND HYDROGEN

BILLION TONS

CARBON	Carbonate Rocks	21,000,000
	Recoverable Fossil Fuels	3,100
HYDROGEN	Water in the Oceans	150,000,000
	Recoverable Fossil Fuels	270

Figure 1.

function of providing both energy and raw materials for the chemicals and other industries. At times these demands may well conflict. This coming summer, for example, could see strong competition between the motorist and the chemical industry for the light gasoline fraction in Europe.

There are many potential sources of energy. As an indication of the magnitude of energy flows in the earth's system it is interesting to consider the total picture, illustrated in Figure 2. This shows that total world energy consumption in 1972 (5,400 million tons oil equivalent) was only equal to 0.004% of the total solar energy input to the earth.

However, although other energy sources, such as solar and geothermal, have been and will be used in certain places for certain applications, their effect on total energy supplies will be very small for some time to come. For example, the report prepared by the U.S. Atomic Energy Commission for President Nixon, which recommends that some $400 million be spent on solar and geothermal R & D over the period 1975—79, estimated that by 1985 these sources might be contributing only about 3% of the nation's total energy requirements.

Thus the problem with all new energy sources is one of time. It is unlikely that any significant additions to world energy can come from any new or relatively new sources in less than ten years, unless developments are already well under way. As another example, even if nuclear power grows at over 30% per annum for the next eight years its share of world energy in

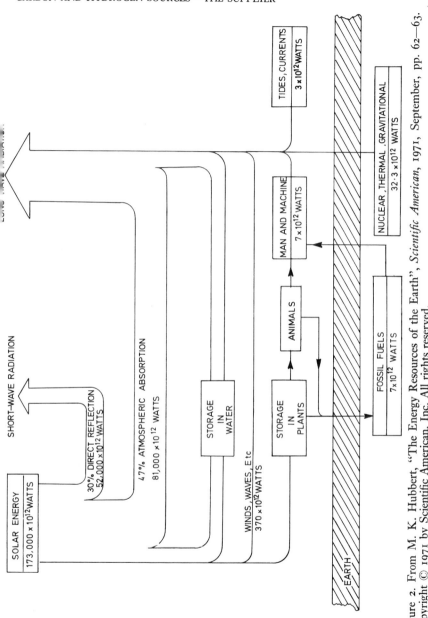

Figure 2. From M. K. Hubbert, "The Energy Resources of the Earth", *Scientific American*, 1971, September, pp. 62—63;

1982 would still only be 5% compared with about 2% today (on an energy input basis).

The events of, and since, October last have swept away a number of familiar landmarks in the energy, particularly the oil, scene. It was difficult even before then however, to make predictions for we could foresee then the emergence of a number of varied problems including a large build up of Arab funds, balance of payments difficulties, problems of the environment and the massive investment requirements, all of which tended to stand in the way of the orderly development of energy sources to meet the world's

PROVEN   OIL   AND   GAS   RESERVES

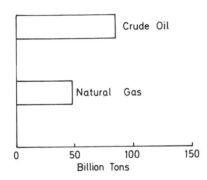

Figure 3.

needs. Since October, crude oil prices have quadrupled and oil has been used as a political weapon.

Each of these factors will have a bearing on the way in which the future can now be foreseen as far as demand, national security, economy of use, and stimulating the development of other sources are concerned. However, it is clear that, in the main, fossil fuels will have to meet our requirements for the next 10—15 years.

Let us now consider world reserves of fossil fuels. Proved world reserves of oil and gas, shown in Figure 3, total rather less than 130 billion tons. Figure 4 shows the picture, if possible additions of oil and gas are included, and tar sands, oil shale, and coal are added. The total of all these reserves— some 3,400 billion tons of oil equivalent—would meet world energy demand at today's consumption levels for just over 600 years. Coal, of course, is much the most significant of all these sources. Looked at another way, so

WORLD RECOVERABLE HYDROCARBON RESERVES

End 1972 Billion (10^9) Tons Oil Equivalent

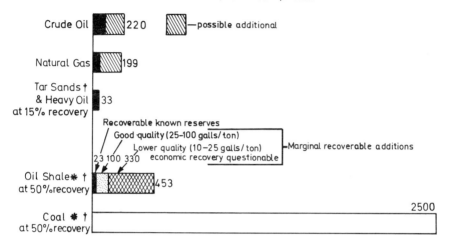

Figure 4.

WORLD PROVEN OIL RESERVES

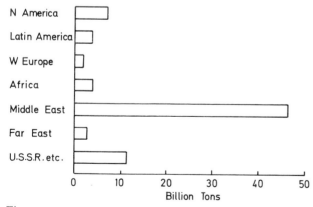

Figure 5.

far we have used up some 16% of total possible recoverable oil reserves, and only about 4% in the case of coal. Hence, for a long time to come, there should be no shortage caused solely by inadequate reserves. Having said that, however, I must immediately qualify it by saying that there are many important and difficult problems to be overcome in the technical, political, environmental, and financial spheres before these sources become available for general and adequate use. The ability of technology to respond to change is considerably slower than the rate at which changes can occur; in spite of the almost 'overnight' quadrupling of oil prices which made the

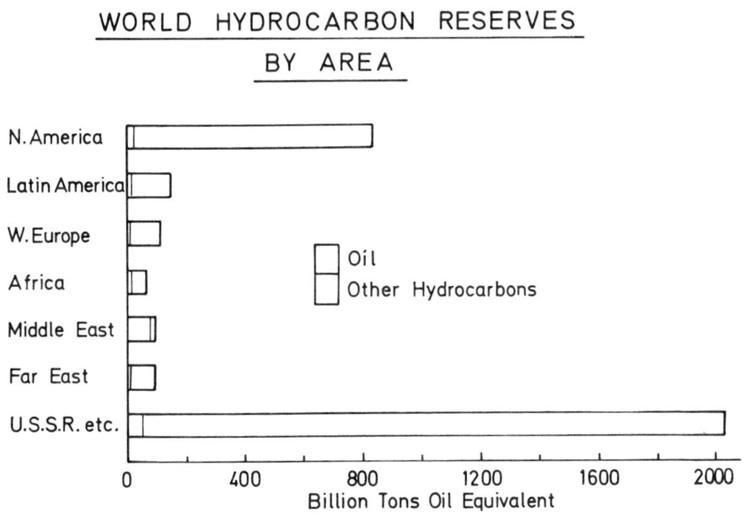

WORLD HYDROCARBON RESERVES
BY AREA

Figure 6.

development of tar sands and oil shale economic, it will be many years before such developments are making any sizeable contribution to supply. However, over a reasonable period, demand at the right price can be met. The problem is then basically one of timing, price, investment, and politics.

The breakdown of oil reserves by geographical area is shown in Figure 5. The strength of the Middle East as the world's major oil source is clear, at least until alternative fuels to oil can be found which will reduce the dependence of world consuming areas on oil. At the present time, Middle East production accounts for one third of total world requirement and on the basis of the above reserves, of which it holds some 60%, its importance is unlikely to decline during the next decade.

WORLD OIL SUPPLY AND DEMAND 1972

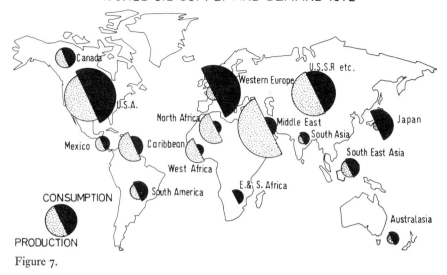

Figure 7.

CAPITAL COSTS OF BASIC ENERGY SOURCES
OIL EQUIVALENT 100,000 bbl/ DAY PRODUCTION

	£ m. (OUTPUT BASIS)	£ m. (INPUT BASIS)
Nuclear power	1,500	500 (75% load factor)
Coal power station	1,200	400 (75% load factor)
Tar sands	350	
Oil shale	250–300	
Coal gasification	250	
North Sea oil	150	
Gas recovery and liquefaction	120	
Coal Mining	15	

Figure 8.

The breakdown of total world recoverable hydrocarbon reserves is shown in Figure 6 and the importance of Russia and North America is quite apparent. It can also be seen that Western Europe and by inference Japan, which is included in the Far East totals, are relatively poorly off for indigenous resources. As far as oil is concerned many of the reserves are located away from centres of consumption, and transportation is an important factor. This is shown in Figure 7.

Costs and time scales of development are very important in relation to energy sources and the conversion of primary to secondary energy. Figure 8 shows a summary of costs for a number of possibilities, demonstrating the high initial capital cost of conversion to secondary energy compared with fuels for direct burning, even from tar sands. It is also relevant to note that as we move into more difficult areas of oil exploration and production, costs escalate tremendously. The cost of a barrel per day of oil production developed in the North Sea or the North Slope of Alaska is of the order of ten times that which we have as an industry experienced in the Middle East during the past decade. A number of estimates have been made of the sums required for future development of known reserves and a figure of some $200 billion during the next ten years would perhaps not be far out of line. In addition to this sum there will, of course, be further demands on a considerable scale for the provision of shipping, refining, and downstream transportation facilities to bring the finished product to the consumer. In the light of current world environmental concern, these facilities themselves will only be achieved at much higher costs than previously.

As far as time scales are concerned, a nuclear power station currently takes between seven and ten years to bring on line. A tar sands project is likely to take about five years; a North Sea discovery about 4—5 years before the oil will be coming ashore and about seven years for a new offshore area following the issue of licences.

As I mentioned earlier, time is also the problem with all new energy sources. One recent estimate for the possible substitution of coal for oil by 1980, in Western Europe, Japan, and the USA was about 3 million barrels per day, that is, only some 6% of forecast oil demand in these areas. Also, it has been estimated that, by the mid '80s, production from Canadian tar sands and U.S. shale deposits might total no more than 2 million barrels per day. It is clear, therefore, that for the next 10—15 years oil will have to be the balancing fuel since the flexibility of supply from other sources is comparatively limited. As we have already seen the main flexibility of oil supply lies in the Middle East and OPEC countries.

In the following paper, Dr. Youle will be considering demand in detail but I would like at this stage to discuss demand in a general sense. For the past 20 years energy demand has correlated closely with GNP. Over the period 1960–72 energy demand grew at 5.7% per annum, with oil playing a dominant role. It is now likely that, in the short term, the demand for oil will be reduced because of slower energy growth. During this period of adjustment, conservation and substitution by coal will play a part. However, discounting these additional factors, world oil demand over the next ten years may now grow at an average rate of between 4 and 5½% per annum; this compares with the pre-October 1973 forecast of just over 6% per

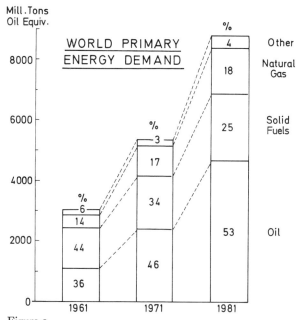

Figure 9.

annum. Historic demand together with a forecast for total world energy demand which was made prior to recent changes in oil price and availability is shown in Figure 9. If all the forces set in train by the recent fourfold price increases materialise, and sufficient action is taken in energy conservation and oil substitution, then the rate of growth of demand may be significantly reduced. For example, by 1981 Western European energy demand may have been reduced by the equivalent of 200 million tons of oil, *i.e.* some 12%, but even then demand will have grown by nearly 20% above 1973 levels.

I would now like to consider the supply/demand situation for oil. As we have seen, current proven oil reserves are about 600 billion barrels or about 85 billion tons oil equivalent. Total world oil discoveries related back to the year of discovery are shown in Figure 10. It is clear that large fields have a disproportionate influence on the total, particularly in Venezuela in 1917, in N. Iraq in 1927, in Kuwait and S. Iran in 1938, in Saudi Arabia in 1948, and

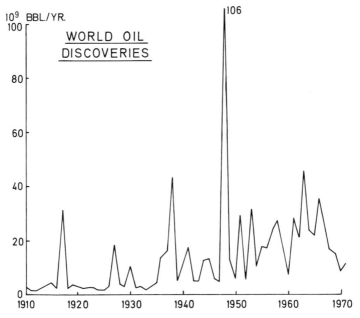

Figure 10.

several large discoveries in 1963—65 in Saudi Arabia and Abu Dhabi. During this period the annual discovery rate was 18 billion barrels. It is not easy from this figure to discern what will be the likely level of future discoveries but, allowing for the fact that much more effort is being put into exploration and that offshore development is relatively new, it is possible that including discoveries of Eastern Bloc oil, total world discoveries might remain at this average rate.

However, this may prove to be an optimistic assumption. The last 25 years has seen the discovery of three-quarters of the oil currently known in

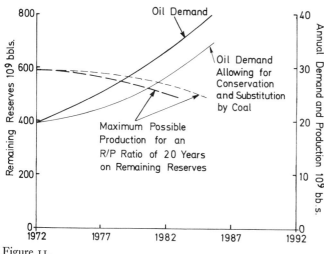

Figure 11.

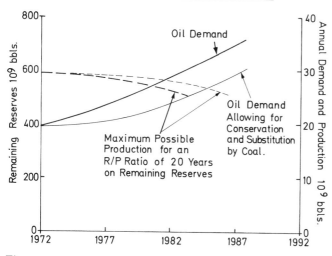

Figure 12.

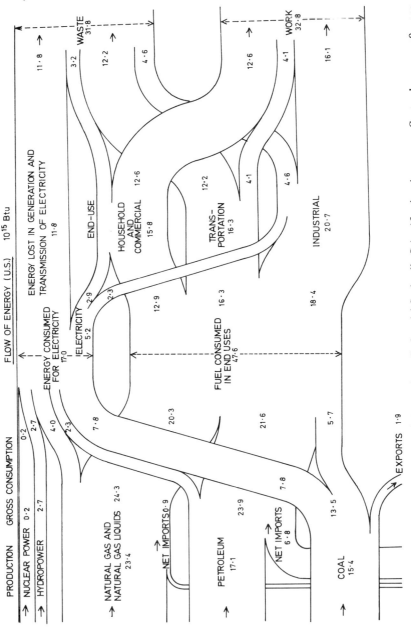

Figure 13. From E. Cook, "The Flow of Energy in an Industrial Society", *Scientific American*, 1971, September, pp. 138—139.

the Middle East and there is no other sedimentary basin remotely similar to the Middle East elsewhere in the world. On the other hand, the recent price increase will clearly have the effect of increasing the attraction of exploration for oil, although as with other resources, exploration effort cannot be increased significantly overnight.

The oil industry feels that it is necessary to have an oil reserves-to-production ratio of 20 years to allow for the lead time required to find new fields and develop existing ones. As we have seen, the discovery rate is 18

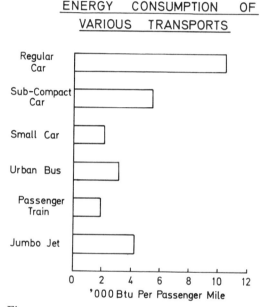

Figure 14.

billion barrels per annum and current production is 20 billion barrels per annum and increasing. On this basis, unless the oil industry is more success-ful in finding new oil in the future than it has been in the past, reserves will decline. I refer to this point, when reserves fall below the 20-year level which we consider to be necessary, as the cross-over date.

As we noted before, current forecasts for oil demand are in the range 4 to $5\frac{1}{2}\%$ per annum. In Figures 11 and 12, world oil demand has been plotted together with potential world oil production, given existing world reserves of about 600 billion barrels, and assuming a continuing annual discovery of 18 billion barrels and an R/P ratio of 20. For a growth rate of oil demand of $5\frac{1}{2}\%$ (Figure 11) it can be seen that the cross-over date is 1978—79 whilst for a growth rate of 4% (Figure 12) it is 1980—81. The cross-over dates can be postponed by 3—4 years by additional reductions in oil demand by energy conservation and substitution, but conversely a lower discovery rate than 18 billion barrels per year would have the reverse effect.

These figures illustrate the simple point that, unless oil consumption is equal to or less than the rate of discovery of new reserves, at some time potential consumption will exceed supply. As can be seen current reserves, and the fact that the world reserve/production ratio is about 30 at the moment, provide something of a buffer for a few years only. Unless prompt action is taken in developing alternative sources of supply, the situation will turn sharply against us from the end of the decade.

While discussing possible overall demand growth, I mentioned conservation and substitution. On energy conservation I think it is relevant to consider current overall efficiency of the use of energy. Figure 13 shows energy flows in the USA in 1970. Only just over 50% of total energy appears as useful work; of the losses, power generation accounts for 37% and transport 38%. Looked at another way, the overall efficiency of electricity produced and distributed was just over 30%, transport 25%, and industry 78%. In electricity generation, for example, the use of combined high-temperature gas turbine–steam turbine cycles, on which the USA may spend over $300 million on R & D over the next six years, and the use of warm water from power station steam turbine exhaust systems for local space heating, could have an appreciable effect.

For transport, Figure 14 shows that in the USA the most significant savings would be made by reducing the size of the car. The use of public transport would help; particularly since, for example, trains should be compared with medium-size cars suitable for long distances, but this will require both a change of attitudes and the provision of effective public transport systems.

Transport is probably a good example of an area where demand is non-price-elastic in the short term, but in the long term some response is both possible and likely.

It has also been suggested that 'anti-consumerism' may develop and thus reduce demand for consumables, such as, for example, new cars. Apart from the variation of petrol consumption with size, it is relevant to consider the total energy requirements for cars. For example in the USA in 1970 consumption of energy for cars (assuming a 'life-time' of 122,000 miles) was as follows:

	%
Petrol consumption	61
Petrol refining and petrol sales	15
Oil consumption, *etc.*	1
Car manufacture and sales	7
Repair and maintenance	3
Parking, *etc.*	3
Tyre-manufacture and sales	1
Insurance	2
Road construction	7
	100

Thus, although a switch to smaller cars would reduce the energy required for manufacture and selling, the most significant effect would be through reduced petrol consumption. In fact it would be argued that since new vehicles are needed to reduce consumption, long life/use of existing vehicles would, in a way, be detrimental.

Let us turn now to the changes which have occurred in recent months and their impact on future developments.

1. *Investment Requirements*

The investment required to provide for future energy supplies will be very considerable. A pre-crisis estimate of the funds required by the oil industry over the next 15 years was one trillion (10^{12}) U.S. Dollars. Although the effect of the recent price increases will be to reduce demand, and hence the investment required in, for example refining may be less, the acceleration of the development of alternative sources will be a counterbalancing factor. There is no doubt that the recent prices increases will have a significant effect on the international economic and monetary systems. The income of the producer countries in 1974 will probably be about four times higher than might have been expected some six months ago. In round terms a previously predicted income of around $30 billion could now be of the order of $130 billion. Conversely on the consumer side, the UK might now incur in 1974 a cost for oil imports of some £3800 million, which represents an increase of some £2600 million over pre-October 1973 prices. Recycling of producer government surplus funds will be essential and consumer governments must, at least in the short term, borrow rather than try to restore financial equilibrium. The result of the latter course would be a significant downturn in economic activity, which although it would have the effect of reducing the demand for oil, would make it even more difficult for consumer governments to pay for their oil as the whole economic cycle goes into decline. These matters can surely be solved only in an international forum, where all the significant aspects of world trade and finance are taken into account.

2. Controls

Another important aspect is consumer government controls. The problems of inflation and balance of payments mean that some form of control over the prices of all basic commodities may be felt to be necessary. However, as I noted before, significant investment will be necessary if future energy requirements are to be met. Such investment can be provided by industry and/or governments. Bearing in mind the international nature of energy (many centres of demand are relatively poorly off for reserves) it is for many reasons preferable for the necessary cash flow to be generated by the oil and energy companies rather than governments. Thus prices must be such as to allow industry to provide the facilities needed to meet our future energy requirements, and adequate allowance must be made for the effects of inflation.

3. The Environment

As important as any factor in the total equation is, of course, the need for each country to come to terms with the conflicting requirements of total purity of the environment and the reality of its energy needs. An example of this is in the USA where the effects of the legislation regarding car exhausts has been to increase petrol consumption at a time when the USA is short of gasoline.

The oil industry has never been more aware of its role of ensuring that a reasonable balance is struck between its responsibilities to the environment and to industrial progress, but unless governments and the public at large equally recognise the dangers of environmental procrastination, we may well lose out in the time scale required for energy developments.

4. Chemicals

In view of what I said about the time needed to introduce new technology, the chemical industry will have to continue to use oil-based feedstocks at least in the 10—15 year period which we have been considering. Looking further ahead, although in certain special circumstances alcohol produced by fermentation and from coal can be economic raw materials, it seems likely that in the main the petrochemical industry will continue to be based on oil. Certainly, chemicals represent a premium market; in general terms it is easier and more economical to substitute other fuels for the oil which is burnt for power and heat generation, rather than switch the feedstock for chemicals to other sources. In the UK, the North Sea should provide tremendous opportunity for the UK chemical industry by supplying a domestic source of both gas and oil. However, this need not necessarily rule out the need for evaluating various oil fractions. The chemical industry should start to give some consideration to the possible use of a wider range of oil and gas feedstocks.

Summary

I have discussed energy sources and availability and the forward oil supply/demand position. There is no denying that there are grave problems but these are not insoluble if appropriate action is taken now. Some of the most important will be the decisions taken in respect of world trade, inflation, and balance of payments. Equally there is a need for a rapid start in developing new sources of energy and in improving the efficiency of the use of energy and in substituting other fuels (e.g. coal) for oil where the efficiency is low and the premium properties of oil are not really used, e.g. power generation. The transport sector requires analysis and there is a need to develop more economic systems; greater interest may be shown in public transport systems. Higher prices will certainly play a part in moderating the increase in demand, but as yet is is not clear by how much. The era of 'cheap' energy has passed, but on the other hand, supplies can and will be available at the right price. The chemical industry, with its advanced technology and the high added value of its products, should be able to demonstrate its ability to survive, despite the pressures on all sides which the future may bring.

Carbon and Hydrogen Sources—the User

by P. V. Youle and J. R. Stammers; Imperial Chemical
Industries Limited

Introduction

BEFORE starting on our official brief, about carbon and hydrogen and hydrogen sources from the user's point of view, we must mention two other important resources. Besides hydrocarbons, the chemical industry depends crucially on these two other sources, which are related to each other.

The first is money. The business of the chemical industry depends on building plants. Changes in the availability of carbon–hydrogen sources may cause us to want to build new plants. But there may not be enough money for all the plants we want to build, along with all the new power stations, oil installations, battleships, schools, roads, and so on. Important debates should take place on priorities and we should take part in those debates.

Wealth is, of course, what we create by hard work and it is therefore closely linked to our second resource—people. Of vital importance to industry is the creative talent of all the people it employs.

The chemical industry also interacts with those outside it and they too are a resource. In the face of scarcer and more expensive carbon and hydrogen feedstocks the chemical industry might want to recycle or re-use certain products. If the industry had such wishes, the plans would need full constructive discussion with the public at large, or the plans would not work.

We do not believe that we can answer our questions about the use of carbon–hydrogen sources solely from inside our chemical technology system. We have to relate our findings to the needs of the broader human system of which the chemical industry is a part. We believe an understanding of the existence of hierarchies of interconnecting systems is fundamental to our whole discussion, and crops up again and again. Incidentally, our discussion is based on an assumption of reasonable stability in the broader human system—no major revolutions and no atomic wars.

Let us now turn to the picture of carbon and hydrogen use as seen from inside the chemical technology system. What we should like to do is to build up as logical a picture as possible, starting with human needs, and ending with a statement on how the chemical industry is going to use carbon–hydrogen sources to satisfy those human needs.

As you well know, though, the chemical industry is not the only one making demands on carbon–hydrogen sources. We compete for them with the transportation and energy industries. Therefore we shall have to allow for their demands and say something about their future situation as well as about the future of the chemical industry.

Thus we hope to look at the chemical industry as part of the total human system so that we can take an overview of what the chemical industry uses carbon–hydrogen sources for and what sources are available to the chemical industry. By comparing demand and supply we hope we can agree on a strategy for a change programme. Then we should like to discuss some of the more detailed tactics of that change programme. Finally, we will try to say broadly what changes all this might lead to, in the three industries which depends crucially on carbon–hydrogen sources: the transportation, energy, and chemical industries.

Demand

Uses of Carbon and Hydrogen in General

Carbon and hydrogen are two of the most essential elements in our life today. To illustrate this, let us look at some of the basic human needs. Table 1 shows examples of these and indicates whether or not carbon and hydrogen are involved in their chemical makeup.

Table 1: Some Human Needs and their Carbon/Hydrogen Content

Need	Examples	Are C/H involved in chemical makeup?
Food	Carbohydrate	yes
	Protein	yes
	Fat	yes
Clothing	Wool	yes
	Cotton	yes
	Synthetic fibres	yes
Shelter	Wood	yes
	Plastics	yes
	Metals	no

Table 2: Some Sources of Energy and their
Carbon/Hydrogen Content

Examples	Are C/H involved in chemical makeup?
Solar	
Coal	yes
Oil	yes
Natural gas	yes
Nuclear	no
Geothermal	no
Gravitational	no
Tidal	no
Electrical	no

There are other needs—for warmth and transportation (see Table 2). Carbon and hydrogen are certainly present in the most commonly used heat sources. But here the chemical makeup of the molecule is not important in itself, it is the energy which it is capable of releasing in which we are interested. And hydrocarbon fuels just happen to be convenient sources of energy which have been readily available in reasonable quantities for a long time at— until recently—a low price. A look at world consumption figures (Figure 1)

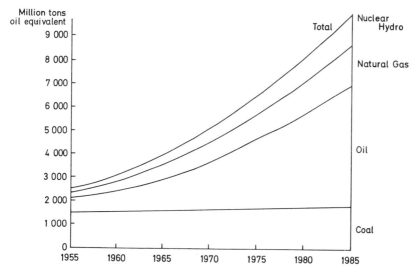

Figure 1. World energy consumption.

Table 3: Major End Uses of Oil Consumption in 1972[1]

Country	Electricity generating	Road transport	Chemical feedstock	Iron & Steel	Other industry	Domestic & commercial heating & lighting	Other uses
			Percentages				
U.S.	9	45	5	1	4	19	18
Japan	18	15	10	n.a.	30	8	19
W. Germany	8	13	$5\frac{1}{2}$	$3\frac{1}{2}$	24	34	12
France	$12\frac{1}{2}$	20	4	n.a.	30	25	$3\frac{1}{2}$
Italy	20	18	$7\frac{1}{2}$	n.a.	24	$24\frac{1}{2}$	16
Belgium	$19\frac{1}{2}$	15	$6\frac{1}{2}$	$2\frac{1}{2}$	26	26	$4\frac{1}{2}$
Holland	$10\frac{1}{2}$	19	$18\frac{1}{2}$	$2\frac{1}{2}$	13	27	$10\frac{1}{2}$
U.K.	21	$21\frac{1}{2}$	6	5	24	$10\frac{1}{2}$	12

Sources: OECD Oil Statistics, 1971 for US and Japan
EEC Energy Statistics, 1972 for European Countries
[1]NOTE: 1971 for U.S. and Japan

shows how the proportion of our energy needs met by oil has increased over the past twenty years.

So we see that hydrocarbon compounds have two main uses: to make other compounds containing carbon and hydrogen, and as sources of energy. Table 3 shows a more detailed breakdown of oil consumption in 1972 in the U.S., Japan, and Western Europe. You can see how the demand pattern differs in the various countries—in the U.S., for instance, 45% of oil consumed is used in road transportation whereas in West Germany the figure is 13%. This is reflected in the processing used in refineries in the individual countries. Table 4 compares the yield pattern in the U.S., and Western European refineries in 1972; again, notice the high proportion

Table 4: Yields on Crude in Petroleum Refineries
1972
(% by weight)

	U.S.A.	W. Europe
Gasolines	43	17
Middle distillates	29	32
Fuel oil	6	37
Other products (including refinery fuel and loss)	22	14

Source: B.P.

Table 5: World Oil Con-
sumption by Area 1972

Area	%
U.S.	30
W. Europe	27
Japan	9
Rest of world	34

Source: B.P.

of gasoline manufactured in the U.S. And, of course, the U.S. is a very large oil user (*c.f.* Table 5). So, on a world basis, we can say, very approximately, 25% oil is used for gasoline, 69% for other power uses, and 6% for chemical manufacture.

Use of Carbon and Hydrogen in Chemicals Manufacture

Let us consider in more detail what we do with this 6%. The gross output of the chemicals and allied industries in the U.K. in 1971 is shown in Table 6. Most of these involve the use of carbon and hydrogen in some way. And 90% of organic chemicals production is made from petroleum. The detail of this Table is not the important point; running our eyes down the headings —pharmaceutical preparations, paint, fertilisers, and so on—we realise how completely our lives depend on the output of the chemical industry.

Table 6: Gross Output of the Chemicals and Allied Industries Sector in the U.K., 1971

	£million
Inorganic chemicals	399.3
Organic chemicals	501.1
Other general chemicals	387.1
Pharmaceutical preparations	497.6
Toilet preparations	169.1
Paint	221.7
Soap and detergents	240.7
Synthetic resins and plastics materials and synthetic rubber	517.4
Dyestuffs and pigments	191.7
Fertilisers	279.0
Other chemical industries	522.6
Man-made fibres[1]	365.8
TOTAL	4293.1

[1] 1970 data
Source: Provisional results of Census of Production

Table 7 gives a clearer idea of the tonnages of chemical intermediates involved.

Not so many years ago, I.C.I., like the rest of industry, used coal. The switch to oil was carried out with efficiency and much to everyone's benefit—we would not have these necessary tonnages otherwise. Now we are seeing a second major change. This time the change is from the carefree, abandoned, use of oil to the careful, enforced, husbanding of carbon–hydrogen sources. This second change is going to be even more socially important than the first.

Table 7: U.K. Production of Some Organic Chemicals 1972

	Thousand metric tons
Ethylene	1121.8
Propylene	542.1
Butadiene	200.5
Benzene	556.2
Toluene	321.8
Formaldehyde	121.8
Propyl alcohols	165.8
Ethylene oxide	198.1
Acetone	159.2
Phthalates	101.2
Synthetic rubbers	272.4
Amino plastics	137.8
Polystyrene	160.5
Polyolefins	471.3
Poly(vinyl chloride)	314.4

Source: Business Monitors, DTI

Availability of Carbon and Hydrogen

The available forms of carbon and hydrogen range from simple inorganic molecules—like water and carbon dioxide—to the relatively complicated organic ones found in vegetable matter and petroleum.

Mr Walters has already shown how solar energy converts CO_2, by the photosynthesis cycle, into vegetable matter and fossil fuels. And he has also discussed in detail the fossil fuel reserves situation. But it may be helpful to list (Table 8) the world reserves, as we presently understand them, in terms of carbon and hydrogen. The carbon and hydrogen in vegetable matter can be regarded as being part of the atmospheric CO_2 and water sections. Two per cent of the atmospheric CO_2 (about 110×10^9 tons carbon) is

Table 8: Approximate World Reserves of Carbon
and Hydrogen
(10^9 tons)

Source	C content	H content
Water		140,000,000
Carbonate rock	25,000,000	
Atmospheric CO_2	50,000	
Coal	2,800	110
Oil shale	360	45
Tar sands	25	3
Crude oil	185	25
Natural gas	150	50

used in photosynthesis each year; but CO_2 is returned to the atmosphere by plant respiration and decay and fossil fuel combustion. The latter is causing an increase in the amount of CO_2 in the atmosphere at the moment (*c.f.* Table 9) but the figures given in Table 8 give a reasonable idea of our present understanding of the relative sizes of reserves.

Table 9: CO_2 Produced by Fossil Fuel Combustion
1950—1967

Year	1950	1955	1960	1967
Billion tons CO_2	6.4	7.9	10.5	13.4

Comparison of Demand and Availability of Carbon and Hydrogen

How does our projected demand for carbon and hydrogen match this picture of availability?

Assuming for the moment that the proportion of the total energy provided by nuclear and 'hydro' means will not increase and that the demand for fossil hydrocarbons will increase at its present rate, then we can make an approximate plot (Figure 2) of the future demand for carbon, in both energy and chemical feedstocks. This rate of growth is, of course, unreal—it assumes, for instance, that it will take the U.S. family the same amount of time to become a four-car family as it took to become a two-car family, a rate of growth difficult to sustain. So the argument is rather similar to the calculations quoted a few years ago which proved that there would be more scientists than people in about the year 2000. But nevertheless the plot provides a useful yardstick against which to measure reserves.

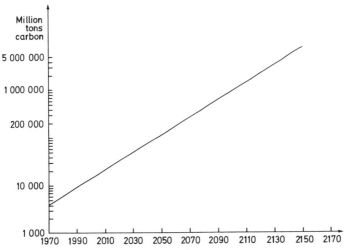

Figure 2. Future world carbon demand.

If, therefore, we show the carbon in fossil fuel reserves, as we presently understand them, on this graph (Figure 3) we shall see that the recoverable reserves would be used up within the next hundred years. And then we would start on CO_2 and carbonate rock. The wide publication of discussions similar

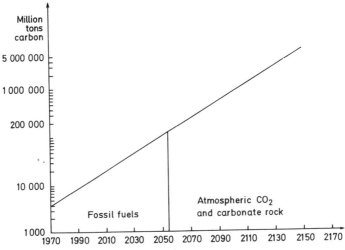

Figure 3. Future world carbon supply.

to the above, together with the recent moves by the oil producers, have caused all consumers to examine their use of fossil fuels carefully and consider how their needs might be met in the future. We must now use these graphs to make some generalisations to govern our change programme.

We can draw two conclusions. First, in the chemical industry oil will be with us for a long time yet, especially if we can, as we must, bring down the slope of the total consumption line and also reduce the proportion of oil used for non-chemical purposes. As Mendeleev said 'Oil is much too valuable to waste it as a fuel'. That might be our first conclusion. Our second? The chemical industry must start laying down sensible research programmes to conserve oil and to explore alternative sources of carbon and hydrogen and we must encourage our colleagues in the transportation and energy industries to take even more emphatic steps in their sensible research programmes.

Necessary Change Programme

Factors Governing Rate and Direction of Change

Exactly what changes are possible in our energy and petrochemicals feedstocks within the next few decades? To try to make this section as useful as possible we have looked over the shoulders of a lot of our colleagues. They have allowed us to use some of what we saw. But both they and we want to emphasise how tentative it still is.

The direction we take in the future will be governed by a number of inter-related factors: availability and cost of materials; availability and cost of technology; availability and cost of manpower; availability of money; acceptance by the market. The last item is probably the most important. People have determined our present life styles. Some factors governing the marketing of manufactured products are listed in Table 10. But this is only part of the story. You must also be willing to spend money on research if

Table 10: Some Factors Governing Marketing of Products

Price of product on market
Price of competing products on market
Price of substitute products on market
Properties of product
Properties of competing products
Properties of substitute products
Need for product

you want new technology to be developed, and to spend money on innovation and new plant if you want new ideas to be brought on to the market. Suitable manpower, too, must be in the right place for new plant to be built and for technology to be developed. And very likely this will mean training, and retraining.

In making our forecasts we must be realistic about time. It takes time to build a new chemical plant—say, 3 years. It takes much longer—say 12—15 years—to build a nuclear power plant. And it takes 3—5 years, say, to develop new processes. It takes time, too, to innovate. A recent discussion of the process of innovation shows that only half of the sixteen major chemical innovations of the 20th Century show a time interval of less than 10 years between conception and innovation (see Table 11). You might argue about

Table 11: Time Interval between Inventions and Innovations

Innovation	Date of invention	Date of innovation	Interval in years
Nylon	1928	1939	11
Silicones	1904	1943	39
Teflon	1941	1943	2
Penicillins	1928	1944	16
Power steering	1925	1931	6
Radar	1922	1935	13
Television	1919	1941	22
Jet engine	1929	1943	14

Source: A Brown, Chemtech, 1973, December

some of those dates (TV was introduced before the 1940s in the U.K.). But what is quite clear is that anything not yet invented is for the mid-eighties and not much sooner. Our suggestion would be to take advantage of the necessary time interval to do a good job on plant design. The consequences of getting a bad design will be increasingly serious, because the very factors we are discussing today will operate to push up the cost of new plant.

If fuel prices go up, the cost of the materials we use to build our plant will increase. All industries use inputs from other industries, which are themselves users of energy and are also indirect users of hydrocarbons. The *Financial Times* published an interesting set of figures early this year (Table 12) showing the total value of fuel used directly and indirectly per £1,000 of final output in each industry in the U.K. Look down the list at the in-

Table 12: Leading Energy Consuming Industries in U.K.
Value of total fuel used directly and indirectly per £1000 of final output

Industry	£
Electricity	237.1
Cement	206.6
Semifinished & finished steel products	138.6
Fertilisers	132.6
General chemicals	118.2
Iron castings	114.1
Bricks, fireclay refractory goods	109.8
Misc. building materials	84.5
Cans and metal boxes	81.4
Stones, slate, etc.	79.7

Source: Financial Times, 1974, 17 Jan., 28

dustries from which new chemical plants are derived. You will see abundant evidence of a trend upwards in new plant costs, especially if you imagine wages rising in proportion. Thus it may be wise for the chemical industry to extend slightly the time from invention to commercialisation. We should also make sure that designers use a systems approach to optimise plant design.

We have to consider, too, the availability of materials. We have discussed the hydrocarbon reserves situation. But we have left out of our considerations so far the availability of non-hydrocarbon energy sources. Mr Walters has dealt with this, but it may help to list the available sources here (Table 13).

To summarise, then, to build new plant, we need materials, suitable technology, manpower and money at the same place at the same time. To develop new processes and technology, we need suitable manpower and

Table 13: Available Non-hydrocarbon Energy Sources

Source	Potential (10^9 watts)
Solar	177,000,000 ($10^5 \times$ present installed electric power capacity)
Water	2,900
Tidal	64
Geothermal	10
Nuclear	very large if breeder and fusion reactors used

Source: M. K. Hubbert, Scientific American, 1971 Sept.

research—at the same place at the same time. So we can see that changes in our energy and petrochemicals feedstocks situation cannot take place abruptly. We cannot predict that on such and such a day in such and such a place the situation will be such that all cars will be battery-driven or that all ethanol will be made by fermentation processes. Change will take place by different routes at different rates in different places so that for many decades we shall see a gradually changing spectrum of solutions to our energy and feedstocks problems. However, we should like to discuss some of the changes which we think are most likely to take place and indicate the period at which they will begin to be feasible.

Likely Changes in Transportation
In the transportation area, increasing fuel prices will have little immediate effect. There may be a cutback in gasoline consumption if the price reaches a level which the market will not tolerate: people may tend to increase their use of public transportation or they may move towards the use of smaller cars. In the U.S., where the second family car is often a compact model, this is easily done. In this country where one car per family is more usual and represents the investment of a sizeable proportion of a man's income, such a change is less quickly made.

We must remember, too, that a lot of money is involved when a new production line for the manufacture of cars and aeroplanes is brought onstream and the manufacturers will want to make that particular model for long enough to make the investment pay off. And think what a time it has taken to build Concorde. So widespread introduction of new forms of transportation (such as battery-driven vehicles) cannot be made quickly, even if satisfactory technology has been developed. Anyway, liquid fuels are an extremely convenient source of energy: we shall probably never see a coal-fired aeroplane. But we shall, I believe, see some addition of methanol (possibly made from coal) to the petrol we put in our car engines. And certainly the time will come when it will be feasible economically to make petrol and other hydrocarbons from coal. In those areas of the world with no oil reserves and large coal deposits this is particularly attractive and in South Africa synthetic crude has been made from coal since 1955. The technique is far from new, being based on the Fischer–Tropsch synthesis process developed in Germany during the Second World War. The South African Coal, Oil and Gas Corporation not only makes synthetic petrol but other hydrocarbon products. This could be increasingly important, there and in India. Eventually, of course, we may feel less need to travel so much

Table 14: Likely Changes in Transportation

1974→	Cutback in gasoline consumption if price reaches a level which the market will not tolerate
1974→	Increased use of telecommunications
1976→	Production of petrol from coal, if economic
1976→	Introduction of methanol
2000→	Other methods of transportation

because of the development of sophisticated telecommunications techniques. So we can summarise the likely changes in the transportation field in Table 14.

Likely Changes in Large Stationary Power and Heating Installations
Immediately, of course, there can be little change in this area beyond modifications to existing plant to improve energy recovery. This is no small point—at present, two-thirds of the power produced in electricity generators is lost in the turbine condensers. And this waste heat could well be used in, for instance, district heating systems. But certainly the first moves away from the use of oil should be made by those responsible for power stations and other large stationary power installations. This is unattractive to the power people—a large amount of investment has been made on oil-fired plant. But this is an area where coal, or natural gas (possibly synthetic natural gas made from coal) and the unconventional sources of energy can make the most impact. Provided, of course, we are willing to spend money on converting oil-fired plant to solid-fuel units and invest even more on new plant. Remember that it takes 12—15 years to build a nuclear power station, and reserves of uranium are only sufficient to meet the needs of fission reactors for the next couple of decades; fusion reactors are not likely to be possible

Table 15: Likely Changes in Large Stationary Power and Heating Installations

1974→	Improved energy recovery
1967→	Increase in use of coal and natural gas
1976→	Introduction of synthetic natural gas (made from coal)
1980→	Increase in exploitation of present nuclear power techniques
1980→	Increase in use of geothermal, tidal and water power
2010→	Introduction of nuclear fusion
2050→	Large-scale use of solar energy

until the turn of the century. So nuclear power cannot provide a significant part of our power needs for a long time. In the very long term, of course, we shall move to solar, water, and geothermal power. There is a real need for a systems study of our whole energy policy, world-wide, through the United Nations, to which we must give the necessary authority. So the picture for large power generation will probably be as shown in Table 15.

Likely changes in Chemicals Production
In this concluding section, we should like to share with you some of the ways in which we are trying to analyse the situation. To begin with, in the short and medium term, we believe that hydrogen will go on being linked with carbon in the sources we use. We tend to go on thinking of them occurring together. Later on, hydrogen will need to be seen separately, derived, most likely, by breakdown of water.

The decisions concerning the processes we are using, and those we shall change to, will be governed by the following factors: feedstock availability and costs; feedstock costs; energy costs; capital costs; and process yield. Certain local considerations will apply. In the Middle East, there may be short-term encouragement for the building of large petrochemical units— to such an extent that there might be the danger of overproduction in a few years time. The petrochemical units may be linked to the manufacture of protein by a fermentation process. Local considerations will affect other countries in special ways.

Considering feedstock availability, coal will certainly come into the picture more prominently in parts of the world where it will be relatively cheaper than oil, like South Africa, or India.

Vegetable matter is already being used in parts of the world as a source of ethanol. Developing that theme, we could imagine large areas of intensive agriculture where every leaf and every animal is used in some way. But we believe that oil will remain a key source of carbon and hydrogen for a long time in the chemical industry in the United Kingdom and the Western World.

Change will occur here, but it will not immediately alter all our basic synthetic routes. There will be quick changes to improve efficiency. There will be research programmes on new synthetic routes still based on oil. Then there will be other research programmes that move away from oil.

On process yields and energy costs, certain things can be said at once. Our big complexes are with us and will stay with us. In such complexes, the chemical industry is already constantly striving to improve raw material efficiences and reduce energy costs. Our efforts will intensify. Existing plants will be modified to improve process efficiencies and more energy will be recovered. Then in addition, for example, distillation columns could be designed with more plates and less reflux—taller and slimmer. In oxidation processes, gases could be recovered and used to drive turbines.

On such big complexes we shall have to consider an overall, total, programme of improvement. Separate improvements interact with each other. For example, improving maintenance might involve a computer system but also retraining staff, and perhaps more interstage storage. Management information systems, if improved, could lead to changes in process control systems, and mechanised chemical analyses in the laboratories. Each good idea needs to be seen in context, and to be related to supportive good ideas in related systems, and ultimately to the whole technology of the plant and to the whole team of people who work there.

A concept of optimised production systems is of fundamental importance, especially when we remember the effects of the big electronic computer— both very good and very bad. Much more work is needed on the Systems Engineering of our complex production units. It almost qualifies as a new University subject. It certainly deserves a new emphasis in Chemical Engineering courses.

Developing that concept of Systems Engineering automatically raises the question of a better integration of the oil refinery with the chemical industry. The next changes we shall talk about stem from moves towards this integration. Looking ahead, the decline in consumption of fuel oils and perhaps gasoline will cause the heavier fractions from refineries to become cheaper. Eventually the pattern of refinery operations will change. Already in the U.S., special units have been added on to refineries for the production of chemicals from the heavier fractions.

The literature contains many references to the production of ammonia and olefins from heavy oils. A little further away is the petrochemicals refinery where the whole barrel of oil is used in an optimised way for gasoline and chemicals production. It may seem further away, but already in the U.S., companies like Du Pont have been talking about it.

We should like to dive now into rather more detailed considerations about our present processes, aimed eventually at setting sensible research targets about which processes need most radical change. Colleagues in I.C.I. Petrochemicals and Plastics Divisions have let us borrow some of their provisional calculations about the oil that goes into their products.

Table 16 shows the amount of naphtha going into each of these products

Table 16: Cumulative Energy Consumption for Some
Large Tonnage Organic Chemical Products

Product	Naphtha equivalent (t/t)
Ethylene/propylene/butenes	1.8
Benzene/toluene/xylenes	1.7
Ammonia	1.1
Methanol	1.0
Terephthalic acid	2.1
Vinyl chloride	1.6
Nylon salt	4.0
Polyethylene	2.2
PVC	1.9
Polyester fibre	3.9
Nylon 6 fibre	5.2
Acrylic fibre	6.9

Source: ICI Petrochemicals Division

tonne for tonne, allowing both for direct chemical usage and also for indirect usage for energy generation. It must be said at once that a lot of work is needed to refine such figures. They depend on local conditions and many assumptions. They combine both chemical efficiencies and energy efficiencies. Getting agreement on the figures between different parts of an organisation may not be easy. As to what practical conclusions the figures lead to, one must be cautious, until a lot more data have been accumulated. But the figures do give some idea of the areas most vulnerable to increased energy costs. They do need to be used in conjunction with other factors: for instance, the energy required for further reactions, the energy required to make substitute materials, and the quantity used in the final, marketable product.

The cumulative energy requirements for the manufacture of acetylene from naphtha are high: 4.3 t naphtha equivalent per t acetylene. This is very much higher than the amount used in ethylene production, 1.8 t naphtha equivalent. Therefore this suggests that oil-based acetylene routes to intermediate products such as vinyl chloride cannot compete with ethylene routes—unless a lot of this energy is recovered in subsequent processing, or unless a more efficient acetylene process can be involved.

We can see, too, that nylon fibre requires considerably more energy input than polyester fibre. This would lead us to suppose that, as energy prices rise, polyester would take over a larger share of the market. But we must remember that the specialist uses are unlikely to be touched. For instance,

Table 17: Energy Dependence of Unfabricated Competing Materials

Material	Density	Cumulative tonnes of oil equivalent per tonne of output	Cumulative equivalent kcal/in³
Aluminium ingot[1]	2.7	5.6	2600
Steel billet	7.8	1.0	1340
Tin plate	7.8	1.25	1680
Copper billet	8.9	1.2	1840
Glass bottles[2]	2.4	0.45	186
Paper and board		1.4	
Cement, dry		0.3	
Polystyrene	1.07	3.18	585
HD polyethylene	0.96	2.33	385
LD polyethylene	0.92	2.24	360
Polypropylene	0.90	2.55	390
PVC	1.38	1.95	465

[1] Assumes 5% recycled
[2] Provisional data
Source: ICI Plastics Division

sales of nylon stockings will surely remain unaltered. And in carpet manufacture things are unlikely to change: Terylene is not suitable and the only material which can compete with nylon in this market—acrylic fibre— has itself a very high energy input. The nylon shirt, however, is in competition with a Terylene and cotton mix. But we must remember that the cost

Table 18: Immediate and Medium-term Changes in Chemical Industry

1974→	Improved process efficiency
1974→	Improved energy recovery
1976→	Improved process control
1976→	Introduction of processes, already developed, for use of heavy oil fractions
1976→	Return to coal- &/or fermentation-based processes in some geographical areas
1976→	Increased petrochemicals production in oil-producing states
1979→	Introduction of new catalysts
1979→	Introduction of new processes using heavy oil fractions using coal- or fermentation-based methods in some areas
1979→	Introduction of petrochemicals refinery
1979→	?Increased re-use and recycling of materials

38 P. V. YOULE AND J. R. STAMMERS

of the fibre constitutes a very small proportion of the price of the final shirt. So that a fractional rise here need not alter the market pattern.

The energy required to make substitute materials must also be considered. This is illustrated in Table 17.

The middle column in Table 17 is roughly comparable with the data in Table 16—it is in terms of crude oil, rather than naphtha, equivalents, but to convert the latter to crude oil equivalents we need to multiply by 1.05 so the data are almost the same. To interpret this Table properly we need to be aware of factors such as transport costs (cement, for instance, is low in energy input but usually has to be transported considerable distances to its

Table 19: Long-term Situation in the Chemical Industry

Atmospheric CO_2	More efficient fixation of CO_2 by photosynthesis; subsequent use of vegetable matter
	Separation of CO_2 (molecular sieves, *etc.*); subsequent reaction with H_2 to methanol, other C/H compounds
Vegetable Matter	Fermentation→H_2, CH_4, other C/H compounds Pyrolysis→H_2, CO/CO_2; formation of hydrocarbons Production of chemicals (*e.g.* cellulose, lignin, tall oil) from wood
Carbonate Rock Water	$CaCO_3$→CO_2+CaO; subsequent reaction with H_2 Production of H_2 by electrolysis; subsequent reaction with CO_2

place of use) and the actual quantity of material used in the final, marketable product. To take an example, it would appear from examination of the middle column in this Table that paper and board would be less vulnerable than polyethylene to energy price rises. But in fact careful calculations have shown that a paper sack requires about $1\frac{1}{2}$ times the oil (or equivalent) input of an LD polyethylene refuse sack. Similarly, for the size of bottle most commonly on the market, a non-returnable glass bottle requires about $2\frac{1}{2}$ times the oil (or equivalent) input of a PVC bottle. So you can see that the weight of material used in the final product must be taken into consideration when looking at the energy requirements of competing materials.

The next step must be some joint discussion between the chemical industry and the people who will be affected in all the interacting social systems, and in other industries. Only then can sensible actions can be laid down.

We have already featured Tables about the transportation industry and the energy industry. Table 18 tries to do the same for the chemical industry. It summarises what we have said about the effects of the new science of optimised production systems, the effect of changed refinery patterns, and the increasing re-use of materials. For the end of the decade it postulates new catalysts and we would like to believe that then, or only a little later, some of these new catalyst systems will be biochemical.

In the medium-term future will come new processes based on oil for the intermediates we select in our 'sensible' research programme.

Table 19 indicates a series of exceedingly exciting long-term research targets. The key to the situation seems to us to be the thought, that our raw materials are on trust to us, and that we must now think very carefully about the best way to discharge that trust, and to husband the resources. Our hope is that the thinking in pursuance of that aim, done by the chemical industry, will help it to stay efficient and useful, but moreover that our thinking and our actions in this new revolution may be one way in which new and better values are injected into our whole society.

Energy from Electrochemical Reactions

by A. B. Hart; Central Electricity Research Laboratories

As the fraction of nuclear-based electric power increases, storing large quantities of electricity in electrochemical batteries could well become obligatory. And later electrochemical processes may be required to assist the conversion and transmission of nuclear or solar energy, e.g., by converting these primary energies into pure hydrogen through the electrolysis of water, and then making electricity available again in the home or the factory by means of a hydrogen/air fuel cell. Indeed, electrochemical reactions may help in the synthesis of organic materials, transportable liquid fuels, plastics, and food substances, from CO_2 and water, using, nuclear energy or the energy of the sun as plants do.

I shall suggest that electrochemists and those who fund their research and development work must be prepared to come forward with new and interesting options in the energy field.

The Future of Electrical Energy

This paper is based on the conviction that electric energy will continue to be an essential and increasing element in the distribution and use of energy in our society. At present in the U.K. electricity accounts for only 11 or 12 per cent of energy consumed at the point of use. But this proportion is rising steadily: it was 9.3 per cent in 1965.

G. F. Ray of the National Institute of Economic and Social Research estimated recently that electricity consumed would increase by 25 per cent between 1975 and 1980, about 5 per cent a year, while total energy consumed would increase by only 12 per cent. Without attempting to forecast oil prices even in the near future it is surely possible to state with confidence that they will increase with respect to the price of nuclear heat. So, too, will coal prices, both home-produced and imported from more lavish deposits abroad. So there could be an urge to replace oil as the primary fuel with nuclear material, and a continued 5 per cent per annum growth in electricity demand until the 1990s seems, on this basis, to err on the conservative side. A 5 per cent growth would increase the electrical energy consumed from 200 TWh at present to 750 TWh in the year 2000, probably requiring about 150 GW of electric power stations at least two thirds of them nuclear. But far more electricity than this could be called for if environmental pressures

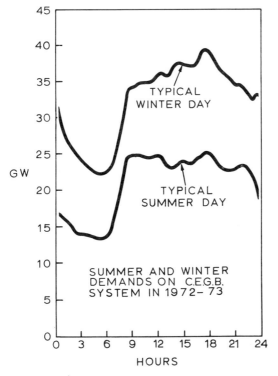

FIGURE 1. Summer and Winter demand on the CEGB System in 1972—3

as well as oil costs decreed that there should be a major shift to nuclear-based electric battery road traction for urban communities. Some fore-casters suggest that 250 MW of capacity could be required by the year 2000. Much will depend on whether suitable traction batteries are available, and this will depend on the electrochemist. Of course, by the year 2000, methanol, prepared from wood or vegetation, linking current solar energy and storable and usable energy, might begin to appear on the scene and this again could call for the intervention of the electrochemist to develop fuel cells which will ensure economical utilisation of the methanol for motor traction. A large increase in the need for nuclear-based energy could heighten interest in large electricity storage batteries to smooth the load on the electrical transmission and distribution system. The familiar daily load curve (Fig. 1) reminds us of the target.

One thinks at once of night-charged, electric battery driven motor vehicles and trains. These would fill up the night-time trough. Already there is a lively interest in electric vehicles for use in urban communities because they would limit the pollution inevitably caused by combustion engines. The electric battery vehicle is silent and free from pollution and it uses its electric power efficiently. In cities the usually rather reduced top speed available with currently available batteries is very little of a handicap. Commonly envisaged are batteries delivering 30—40 kW of power at peak and 3—5 kW at steady load, storing, say, 25 kWh altogether. Charged overnight, 2 million of them with a 6 GW steady output capacity would consume 20 TWh p.a. on charge.

Expansion into the mass market would require large technical and financial innovations—replacement batteries at 'filling stations' and battery costs spread over vehicle life-time: were they to be overcome, the potential 'traction load' by 1990 might be equal to a night-time output of 50 GW of of nuclear capacity!

Another possible future application of the secondary battery is in storing power in the electricity system itself. Storage batteries could be charged with low running cost plant when the demand is low, and the benefit would increase as the fraction of nuclear plant on the system increases. It can be calculated that by 1995 the money available at present-day values for a battery with a moderate overall power to power efficiency of 0.65 could be at least double that permissible for a gas turbine with the same duty. This seems a reasonable target for the electrochemical engineer. There would be other advantages. Batteries could respond extremely rapidly to demand variations and so provide instantly available 'spinning reserve' which currently costs the CEGB about £4M p.a., and could rise to £20M p.a. in the 1990s. Pumped water energy storage, e.g. at Ffestiniog, helps with that now, and the new proposal at Dinorwic, which could give 1500 MW for 5.2 h daily from water stored overnight, could meet needs for some years ahead, but in the long term more capacity for this will be needed and pumped water can only be stored on this scale in mountainous regions with concomitant transmission costs. Load smoothing units also contribute to improved reliability and reduced maintenance of nuclear plant—which are best operated at steady load. Electric batteries have the advantage of siting in small units towards the demand end of the transmission and distribution network, for example in units of 0.5 to 5 MW, and this offers a further benefit in security of supply and in smoothing the load on both the national transmission and local distribution network.

The total amount of battery storage capacity which could be installed would depend not only on its annual capital and maintenance costs but also very sensitively on its power efficiency. One could expect to see up to 10 GW output capacity of suitable battery in 1990—95 and depending on the 'natural smoothing' in the system by then caused by night-time traction battery charging, as much as three times as much ten years later. It is a huge challenge requiring a reorientation of the battery industry.

Storage Batteries

The lead/acid and nickel/iron, nickel/cadmium systems are the lowest cost batteries available now for bulk power storage. The silver/cadmium and silver/zinc batteries are well-known in military applications, but they are too costly to consider for use on a large scale. Lead/acid batteries, as everyone knows, suffer from high specific bulk and weight and a relatively high cost. The zinc/air battery, a development of the air-depolarised Leclanché cell used for railway signalling a hundred years ago, has been the subject of intense research in recent years, and to a lesser extent so have the iron/air, zinc/chlorine, and sodium/sulphur batteries. Other possibilities which have been investigated are the lithium/sulphur and lithium/chlorine and the nickel /zinc batteries.

In the U.K. there are about 75,000 electrically propelled industrial trucks of various kinds, and about 50,000 commercial road vehicles, but few passenger vehicles. The Electricity Council have announced that they have ordered 60 light cars. Their appearance in numbers could change people's attitude very radically. They have limited but acceptable urban performance. They are powered with lead/acid batteries, some S.L.I., some special traction batteries, and it will be useful to dwell a little on these traction batteries. Typically they give 5 h daily discharge after charging overnight and they give 1500 to 2000 charge/discharge cycles in their useful low maintenance life of six to ten years. Their power to power efficiency of 80 per cent is satisfactorily high.

Now a larger 'family' car or delivery vehicle weighing about a ton fully loaded, with a performance suitable for urban as well as suburban use would consume about 0.5 kWh of electric energy per mile of average 25 m.p.h. use. It would need a peak of about 25 kW for moving away and maintaining speed on gentle inclines. A 50 mile range is a reasonable minimum so a 25 kWh battery would be acceptable, but of course twice or three times this range and amount of stored energy would be much preferred. The running cost would clearly be very competitive, *i.e.* (0.5/0.8) × 1.0p per mile, where 1.0p is taken as the price of one kWh of 'overnight' power, or 0.63p per mile compared with 2.0p with an i.c. engine with petrol at early 1974 prices.

But the weight, bulk, and cost of a lead/acid battery even for a 50 mile daily range are inconveniently large for an all-purpose car. At 20–30 Wh per kg, which is what one obtains from existing traction batteries discharged at the 5 h rate, the family car performance would need about a ton weight —the car would be a mobile battery. But lead/acid batteries have good peak performance—about 5—10 times the average. This can be kept up for short 'bursts' without damage to the cells, and technological developments at present in the pipeline are said to offer the possibility of 40—50 Wh/kg over 5h. Married to a light-weight chassis like the Electrical Council car this does offer the possibility of quite reasonable performance and range.

At the risk of offending those in the laboratories of the manufacturers who have done so much to improve the understanding of the operation of the lead electrodes, separators, and electrolytes, one may ask for a redoubled design effort to achieve a much higher proportion of electrochemical utilisation of the lead—it is little more than a third used at present—perhaps by a thinner more porous structure, perhaps by thin-film techniques. And it is perhaps presumptuous, in view of the great deal of excellent research of which the lead/acid battery has already been the subject, to ask for more searching basic studies to link the recent work on the electrochemical reduction and oxidation process of the lead with structural considerations. Perhaps the new methods of looking at surfaces, ellipsometry, very high powered electron microscopy, ESCA and Auger spectroscopy, can tell us more. Of course the same applies to other battery systems. We must remember that in all electrochemical cells the energy transfer mechanism goes on in a very thin layer of matter, that comprising the top layer of the conducting electrode and the electrolyte, a double layer, say 1 nanometer thick: for every m^2 a volume of $10^{-9}m^3$. In this volume we expect a throughput of at least 2000 coulombs a second: about 2 hundredths of an equivalent of electrons, products and reactants, 20,000 equivalents per litre per second; a tall order. Of course, because the volume is smeared out over such a large area the heat and mass diffusion processes are eased, but the final act of exchange does imply a very intense reaction indeed. No wonder the task of ensuring a stable reaction surface through many cycles is such an exacting one.

Now let us turn to the alkaline nickel batteries. They have a relatively small weight advantage on lead/acid. This is partly because of the lower voltage of about 1.2 V. Nickel/iron gives about 30 Wh/kg for a 5 h discharge period, nickel/cadmium about the same. They have a much lower peak power density than lead/acid. Nickel/iron has a longer cycle life—about 3000 cycles—but the power to power efficiency of 60 to 65 per cent is low. This is apparently due to the iron electrode and there is scope for further research and development on this point. Indeed, Russian workers are said to prefer the development prospects of nickel/iron to lead/acid for traction purposes and have set a target of three times the present energy density at half the present cost. They are mounting a massive research programme embodying searching basic studies as well as engineering development.

The nickel/zinc battery fell out of favour because of the tendency of zinc to form dendritic growths on charge—which short the cell. Recent research done mainly with the inspiration of the zinc/air battery is offering understanding of this as well as of the intractable passivation which can overtake the zinc electrode at higher than normally used, but still quite modest, current densities.

Now to the metal/air batteries. They operate in alkaline solution and have the advantage of working with an external reactant, the oxygen from the air. So the stored weight of the nickel oxide is cut by requiring only a thin nickel or indeed carbon electrode with the oxygen taken from the air. But they have to carry fans for circulating the air to the electrodes and a CO_2 scrubbing system. In the zinc/air battery the overall reaction is:—

$$2Zn + O_2 + 4OH^- = 2ZnO_2^- + H_2O \qquad \dots (1)$$

and the electrode reaction of the zinc:—

$$2Zn + 2OH^- = 2ZnO_2^- + H_2O + 4e \qquad \dots (2)$$

is balanced by the reduction of the oxygen:—

$$O_2 + 4H_2O = 4OH^- - 4e \qquad \dots (3)$$

Because of the low electrochemical equivalent weight of zinc a high specific power has been expected—about 140 Wh/kg. But because of the limited peak power performance of the oxygen electrode—which depends upon the gas diffusing freely to the electrode surface—and the low efficiency of the same electrode which confines the power to power efficiency of the battery to about 50 per cent the zinc/air battery has not been successful in engineering development. In the U.S., General Atomics made an impressive device with pumps to circulate the electrolyte and store ZnO separated from the $ZnO_2{}^-$ solution outside of the cell, and Japanese workers also persist in their optimism that a similar device can be made for an electric vehicle. There may be further scope for fundamental and design studies of both the zinc and oxygen electrodes. The oxygen electrode has already been the subject of much research. It suffers the specific difficulty that the oxygen molecule, once it gets to the seat of the desired electrochemical reaction represented by equation (3) in which the molecule must be absorbed on the electrode surface and in contact with the solution, has an alternative choice of reduction path in the reaction:—

$$O_2 + H_2O = OH^- + O_2H^- - 2e \qquad \dots (4)$$

The peroxide ion O_2H^-, though unstable, may build up in the solution. If it attains a concentration of only one hundredth molar the effect would be to lower the voltage in a cell by 0.5 V from the value to be expected if reaction (3) ruled. Fortunately the peroxide species decomposes:—

$$2O_2H^- = 2OH^- + O_2 \qquad \dots (5)$$

at a catalytic surface, and this prevents too high a concentration forming. A surface which absorbs oxygen, as it has to, to be a good oxygen electrode, is usually a good catalyst for peroxide decomposition. But things get worse when a current is flowing. The nature of the surface and the temperature are very important. Nickel, or rather nickel oxide, alone is not much good, *i.e.* it gives a low oxygen electrode voltage at even a small current below 120°C, but is quite good between 180°C and 250°C when the fragility of $HO_2{}^-$ is increased. But silver is useful at room temperature. Much research still remains to be done although new substances like lanthanum cobaltite, which is nearly as good as silver, have been discovered in recent years and enormous strides have already been made towards an understanding of this important electrode.

With a good and long-lasting air electrode the iron/air cell using only cheap and abundant materials would attract interest. Iron is less soluble than zinc and does not grow dendrites and feasible designs have been worked out for 80—200 kW units. Their weight would be only half that of lead/acid for the same vehicle performance. Their cost per kWh of output power will depend very much on the oxygen electrode.

48   A. B. HART

Getting away from the oxygen electrode with its still-present limitations leads one to think of chlorine as the oxidant. Chlorine electrodes using carbon as the conducting surface are well known in industry. One cell design developed by Gould Industries in the U.S. is basically a zinc/chlorine cell with aqueous electrolyte using the innocuous Cl_26H_2O, a solid at room temperature, as the chlorine store. 200 Wh/kg is claimed and an overall cost less than half that of lead/acid, but the device is complex, it requires a refrigerator, and it is probably suitable only for larger powers. Sulphur is a possible oxidant and the molten sodium/sulphur cell, giving 1.8V at reasonable c.d.s, uses a thin sodium ion conducting diaphragm of β-alumina as electrolyte. It has been developed at Ford and G.E. in the U.S., in Japan, and at the British Rail Laboratories and the Electricity Council Research Centre (ECRC) in the U.K. The Japanese and ECRC have built batteries which have driven vehicles on the roads. The ECRC battery uses robust tubular cells which have given 2000 cycles in laboratory testing. The cells operate in the range 250—400°C with sodium on the inside of a β-alumina tube and sulphur mixed with liquid sodium polysulphide on the other side soaked up in the current collector which is a carbon felt. The cells give a high power-to-power efficiency despite resistive losses in the electrolyte and the practical energy density is about 100 Wh/kg, but the cells are still bulky. This limited success has only been achieved after much research into the structure of the β-alumina but more is required especially into its deterioration with use which is said by one worker to be associated with sodium penetration of grain boundaries. The lithium/sulphur battery operating at 260°C was favoured at one time in the U.S. perhaps because it does not require a solid separator since the lithium does not dissolve in the LiCl-KCl-KI electrolyte which is however immobilised with an inert ceramic powder and this mixture has a limited conductivity. Energy densities of 220 Wh/kg and power densities of 110—150 W/kg are predicted and a paper study has been made of such a battery to power a London bus. But it has gone out of favour now. A couple with a most striking power and energy density is lithium/chlorine which would operate at 600°C and give a voltage of 2.8 per cell with molten lithium floating on the molten chloride and the chlorine piped into a carbon electrode. Power densities of 1000 W/kg, *i.e.* twenty times that of the lead/acid, are predicted on the basis of single-cell work and the energy density could be as high as 350 Wh/kg. Containment materials and corrosion, and chlorine handling are the problems here. Rightmire in the U.S. proposed a sealed version with the chlorine trapped in a porous carbon block. It was intended particularly for vehicle propulsion.

No experimental work has been reported on the lithium/fluorine cell despite its high theoretical open circuit voltage of 5.26 V and the very high conductivity of molten lithium/potassium fluoride as the electrolyte at 500°C. Energy and power densities of 2000 Wh/kg and 2000 W/kg are conceivable. The difficulties are apparent but the technology of fluorine handling is now highly developed and one looks forward to hearing of the realisation of a battery, perhaps not for motor cars despite the remarkably small size of the power unit, but for power storage in industrial situations.

Other less dramatic fluoride systems are spoken of. One uses potassium phosphofluoride, KPF_6, dissolved in propylene carbonate as the electrolyte, solid lithium as one electrode and nickel fluoride as the other. With a 6 V open circuit voltage the theoretical energy density is high but the power density is low due to the poor conductance of the electrolyte.

Fuel Cells

I am sure I have said enough to show that battery technology could well have a dramatic future. Much work and much expenditure is necessary to determine whether that is real or illusory. But I have neglected fuel cells! Perhaps this is because I have been looking forward in these recent traumatic times to a world beyond that in which oil and gas are abundant, and nuclear electricity would have to be pressed into service even for road transportation. Yet offering a higher conversion efficiency of gas or oil fuel to electricity, fuel cells may be able to eke out what oil and gas we may still be able to afford.

The basic technical problems of fuel cells have, of course, been conquered, as the success of the Pratt and Whitney hydrogen/oxygen fuel cell in the Apollo manned lunar exploration programme demonstrated for all the world to see. It is a matter of satisfaction that the U.K. work of Mr F.T. Bacon, contributed so much to the success of that effort. The firm of Pratt and Whitney, building on the know-how the space race won them, have since developed a domestic fuel cell accepting a light clean hydrocarbon fuel plus a little water and providing 12.5 kW of a.c. power from two black boxes about as big together as an office desk. The larger box is the d.c/a.c. convertor, for the fuel cell which is a hydrogen/air one, produces only d.c. The hydrogen has to be made by re-forming the hydrocarbon fuel and to resist the CO_2 in the gases (and air), a strong acid is chosen as electrolyte. That device is being consumer-tested. A fully commercial version may follow. Now the same firm propose to build a 26 MW plant to go into the business of quiet, fumeless generation on the scale now reserved for the gas turbine and fulfilling the same purpose, *i.e.* smoothing the load on the power system and assisting with the distribution of power. It is not yet certain that the electrolyte will be acid at 120—150°C as in the 12.5 kW domestic battery; perhaps an ion-exchange membrane as electrolyte will be used at 80°C; (the G.E. company's laboratories have announced a new design of a cell of this construction for which they claim very high power density); perhaps the battery will have a molten carbonate electrolyte at 550—600°C, perhaps a solid zirconia electrolyte which conducts oxygen ions and can be used at 950—1000°C and has been studied at Westinghouse in the U.S. and at Harwell. In any of these cases 35—50% of the total combustion of the fuel will be available as heat, so there are possibilities for using it for domestic heating, or for supplying process heat. The apparent success of the electrochemical engineering of the hydrogen/air fuel cell does not mean that the work of the electrochemist is done. There are still anxieties about the cost of cheap long-lasting electrodes for the low to medium temperature oxygen electrode and there has not yet been any realisation of the higher-temperature versions yielding 'waste' heat at industrially interesting temperatures.

There seems no fundamental obstacle to fuel cell development for vehicle propulsion. It remains to get the technical details right. Hydrogen/oxygen and methanol cells have clocked-up thousands of hours already. One at Thornton, developed by Shell, uses acid electrolyte to avoid trouble with CO_2 and has a relatively low power density. An alkaline cell would be better on this score but there is the carbonate problem: it has been proposed by Alsthom-Exxon though, that the methanol cell could operate on partly carbonated solution.

Note that a hydrogen/air cell, 2.5 kW to 2.5 MW in power rating, would be the perfect complement at the consumer end of a hydrogen distribution system. Gregory in the U.S. and Marchetti of Euratom at Ispra in Italy have advocated such a system and it has been given enthusiastic support by Bockris, who is well-known for his belief in an important future for electrochemistry. The hydrogen would come from nuclear power and water. For domestic use one would have to decide whether to provide each house or flat with a 2—5 kW fuel cell, which would be arranged so as to give 10 to 15 kW of heat for heating purposes, or do everything in a neighbourhood 2.5 MW fuel cell/district heat installation. Note that the hydrogen distribution system competes with the electrical grid. Hydrogen pipelines would be easier to put below ground than high-voltage lines for an equivalent power Furthermore, losses with gas can be made smaller than with electricity. Even more important, perhaps, the pipelines could store appreciable amounts of energy. The present natural gas grid can store about a day's use. Load smoothing should be easy and economical. One can see that getting power away from a massive 20 or 30 GW nuclear complex, perhaps sited remotely or even off-shore as Americans are proposing for their eastern seaboard, might be more convenient in hydrogen pipelines than in electric transmission lines even at 765 or 1300 kV which are possible in the future. But on the debit side there are efficiency losses in hydrogen generation and the cost of plant at either end. At the consumer end the efficiency of conversion of the chemical energy of hydrogen into electricity might be only sixty per cent. But on the other hand that would not matter too much if the remaining part of the chemical energy was made available as useful heat. The only real losses would then be the ten per cent or so of 'parasitic power' for driving circulating pumps and so on. Plant costs? Quite impossible to say until someone has tried to design the device in detail but probably about £250 for a domestic device delivering up to 2 kW of electricity and 2—10 kW of heat.

But at the power station end we would have to decide how to make hydrogen. Large-scale electrolysis is a technological reality now: electrolysers are reliable and last for 20—30 years without trouble. Power conditioning, electrolyser, and 'site' costs proposed by different manufacturers agree fairly well: transformer, recitifier, controls: £10/kW, (£5 if the electrical input is in the form of d.c.), site costs (building, delivery, erection, commissioning): £10/kW; electrolysers, projected somewhat, £40/kW at the 1000 kW level. So the total is £60/kW for the installed plant. Scaling-up might reduce this to £40/kW. These figures are all expressed in terms of the input power. The electric power to hydrogen combustion heat efficiency would be about 0.8, with a 0.9 load factor on the whole plant and with the price of the nuclear power at 0.46p—a P.W.R. operating at 0.9 load factor—then the final cost of hydrogen per kWh of combustion heat would be 0.58p/kWh. This is equivalent to 17.0p/Th at the station or 20p/Th distributed to the domestic household. This compares with 7.0p/Th—the present price of gas; not competitive, but there is no ground for rejecting the idea of the hydrogen economy out of hand.

There are those who enthuse about the chemical/thermal route for making hydrogen from nuclear heat. Marchetti of Euratom's laboratory at Ispra is the leading proponent of the route. The attraction is that one avoids the low, 0.3—0.4 Rankine efficiency of the generation of electricity from nuclear heat *via* the steam turbine. In particular, it permits the use of the high temperature (say 950°C) available from a High Temperature Gas-cooled nuclear reactor.

The chemical route involves two or more chemical reactions into which water is fed and hydrogen and oxygen removed. Decomposing steam with iron to yield hydrogen and then heating the iron oxide to yield oxygen and iron again would be such a process but in this case the temperature for decomposition of iron oxide is too high. But there are many similar though usually more complex reaction series in which nuclear heat, at say 600—1000°C could be used. Unfortunately they need several heat transfer steps, and the overall efficiency is unlikely to be more than that of the electrical route, while the plant costs could easily be larger.

Hydrogen once made by either route could be transformed immediately into methanol or a hydrocarbon by reduction of CO_2 obtained from carbonate rocks or from the air. Then the transportation of the energy could be by pipeline or tankwagon to the fuel cells/heat units secreted about the city or in premises where the desired heat or electricity would be re-created. A long-term target for future electrochemical research must be to secure the reduction of CO_2 at an electrode either by uniting the CO_2 molecule with a freshly released H atom while still adsorbed at the electrode surface, or by another route. The interaction of these species at electrodes has been studied and indications of reduced species found, but so far no product like CO, formaldehyde, or formic acid in noticeable yield.

Finally, I refer generally to the use of electrochemical processes for achieving socially desired results at less cost in energy terms than alternative methods. My reference to hydrogen production illustrated the conflict between a chemical/thermal and an electrochemical method. Another illustration is the manufacture of hydrogen peroxide, an essential intermediate in the manufacture of washing powders and indeed a possible oxygen carrier for future high power density batteries for special purposes where pressurised or liquid oxygen is not suitable. In earlier days hydrogen peroxide was made by way of anodic oxidation of sulphate ions but since the last war large-scale production switched to cyclic oxidation reduction of ethyl anthraquinone or to partial oxidation of propane both processes being much cheaper than the old electrolytic process. Perhaps future research might develop a 'new' electrolytic process which could take advantage of relatively cheap nuclear power. There are many similar processes in an electrochemical industry growing in sophistication.

All these possibilities call for increased activity in basic electrochemical research and in the development of engineering applications to realise the potential of existing basic knowledge.

Acknowledgement

This work was carried out at the Central Electricity Research Laboratories and is published with the permission of the Central Electricity Generating Board.

Water—An Overall Picture

by H. Fish; Director of Scientific Services, Thames Water Authority

THE quantitative and qualitative adequacy of water resources is essential to modern developed society in three ways. First as an essential commodity of life support (next in importance to oxygen); second as a primary raw material of modern organised living, industry, and agriculture; third as an essential part of our environment—covering some two-thirds of the global surface. Only in recent years has the reality of this third environmental necessity been generally appreciated, to complete what might be described as the 'eternal triangle' of society's water affairs.

That society should now accept the relevance of the hitherto missing link in water management is most satisfactory to those few chemists and biologists who, about a quarter of a century ago, spearheaded the first British national attack on water pollution, and later on over-abstraction of water resources, with the support only of anglers and country-loving landowners. It is also most reassuring now to learn that most of the actions one took in practically achieving water conservation, without being able to explain very scientifically why, were theoretically correct.

However, the developed societies as a whole have not yet achieved a great deal in practice in bringing their total management of water resources to the required degree of excellence, although by all accounts the United Kingdom has a considerable lead in this direction. There is a very great deal to be done yet in water management, and in aspects of use of other natural resources which give rise directly or indirectly to water pollution, in terms both of extending our understanding of the microchemistry and biochemistry of water and the implications of this, and in the practical water management action to be taken in consequence of this understanding.

Our current approach to water management is based on the growth concept that water demands will continue to increase rapidly into the foreseeable future. However to this is added the novel, and very important, modification that we must apply the most appropriate technology to ensure that the

55

environmental consequences of meeting this increasing water demand are kept within the levels acceptable to the public at large. It is with the factors involved in this approach, the problems and unknowns involved in pursuing it, and the modifications which seem likely to be needed and made in this approach in the future, that this paper is concerned. For obvious reasons, only the position in England and Wales will be considered.

The Present Position

According to the report of the Water Resources Board 'Water Resources in England and Wales' published by H.M.S.O. late in 1973, the present total daily demand for water in England and Wales amounts to 42 million m³. 14.1 Million m³/day is taken for public supply, of which one-third is taken from upland water sources, about one-third from the lowland reaches of rivers, and the remainder is abstracted from underground aquifers, mainly the chalk. The total daily demand amounts to about twice the daily run-off of water during extreme drought. To meet these demands, reservoirs have been constructed to conserve wet weather yields of water, and about half the total water demand is met from such reservoir storages.

Of the total of water abstracted, about 25 million m³/day is taken for industrial cooling purposes, most of which is returned relatively uncontaminated to resources. The balance of 17 million m³/day is taken for public supply and industrial process purposes (including agriculture). Of the quantity of water taken for public water supply only about two-thirds is used for domestic purposes. The balance is taken for industrial purposes. Industry in total takes about 28 million m³/day of water direct from resources. Thus industry takes about four times as much water as is supplied for domestic purposes.

These uses of water for domestic and industrial purposes result in production of about 7·5 million m³/day of domestic sewage and about 7.5 million m³/day of industrial wastewater for disposal. About 50% of the industrial effluent produced is disposed of, after pretreatment where necessary, along with domestic sewage at sewage works, while the balance is disposed of, again after treatment, direct to rivers and tidal waters.

These discharges of wastewater are of course the main cause of river and sea pollution. Yet where the discharges are made to freshwater rivers, they become available for repeated abstraction and re-use downstream, permitting a very considerable economy to be made in the overall use of water resources. In fact this kind of repeated water re-use along the courses of our lowland rivers reduces the gross demand on water resources from 42.4 million

m^3/day to 30.5 million m^3/day. The matter of water re-use, particularly in public supply, presents the most difficult problems in the present management of water. The re-use is highly desirable in quantitative and economic terms, but is open to the question of whether the quality of the river water, arriving at waterworks abstraction intakes, and the nature and effectiveness of waterworks treatment processes, can be kept adequate in the future to ensure the true wholesomeness of the water put into public supply.

The extent of this problem can best be understood by reference to examples. On the freshwater River Thames, in a dry weather flow of about 2.2 million m^3/day in the lower reaches, about half of this flow consists of domestic sewage effluent and industrial process effluent. From this flow, some 1.5 million m^3/day of water is taken to meet some 70% of London's water supply. The river is in a clean condition, highly used for fisheries and recreation throughout its length, and London's water supply is undoubtedly of wholesome quality.

To manage the river in its present clean state and fit for public supply abstraction, it is estimated that an annual cost of £15 million is incurred in the purification of sewage and industrial effluent to high standards of quality. If the river were not used for public supply, but were still to be kept clean enough to maintain a reasonable degree of fisheries and amenity use, effluent treatment would still cost at least some £10 million annually. In this case the development of the required sources of water, other than from the River Thames, for public supply would cost some £20 million per annum. Thus there is a net financial advantage of £15 million annually between management of the river as a high-class amenity fit for public supply abstraction, and management of the river solely as a reasonable amenity.

In contrast, the River Trent is at present too polluted for use as a source of public supply and its amenity value is very much lower than that of the River Thames. If it could be cleaned up for public supply use, then after taking into account certain differences between the two river systems, not only would the financial advantage accruing be about the same as that currently being realised in the Thames area, but also the amenity value of the Trent would be greatly enhanced. However, these estimated financial advantages of securing maximum re-use of water *via* rivers have to be balanced against the water quality control problems arising, about which more will be said later.

The Future Position

The prognosis of the Water Resources Board is that by the year 2000, the present demands for water will be nearly doubled. Doubling the use of water will of course double the volume of effluents to be disposed of, and while this will favour greater water re-use it will obviously make water pollution problems more severe. However, to minimise the overall costs of providing new supplies to meet expected future demands, and to minimise the overall environmental impact of reservoir construction, a particular strategy of water resources development is favoured by the Water Resources Board. This basically involves the construction of a few large reservoirs, development of freshwater storage in the Dee estuary, and development of groundwater resources, with most of the water made available being piped to the upper reaches of major rivers for abstraction at points of demand downriver. In this way, the clean water flow of the rivers can be augmented to dilute effluent discharges and hence in theory to diminish the water quality problems in those rivers. This strategy appears attractive in many ways, but it seems most unlikely that it alone will meet future requirements in water management, in particular in respect of water quality needs.

There are many problems facing us regarding the future quality of river waters, of public supply abstractions from rivers, and of the sea.

Regarding the control of pollution arising from effluent discharges to rivers, there is not likely to be any great problem arising in maintaining at satisfactory levels, or improving where appropriate, the oxygen balance in river waters. Essentially this depends upon ensuring that readily degradable organic matter in wastewaters is removed to an appropriate degree in sewage and industrial effluent treatment plants, and that industrial wastewaters do not contain toxic or noxious materials which inhibit such purification processes. This costs money, but the technology is well known and is relatively simple in application. There can of course be problems arising from time to time in consequence of a disorganised society giving rise to frequent industrial action in public utilities, and even these can be dealt with in fair degree until near anarchy develops. This control, properly applied, will by and large ensure that rivers are kept, or made, clean enough to support reasonable fish life, and associated flora and fauna, and to meet most amenity requirements. Certainly since the first industrial action in the 'dirty-jobs' dispute in 1970 occurred, the freshwater River Thames has been kept clean despite several periods of industrial action at, or affecting, power stations.

However, this broad control of oxygen balance is not good enough to keep rivers, receiving large volumes of effluent discharge, entirely satisfactory in environmental terms and as sources of public water supply, and to control properly coastal (and perhaps ocean) pollution by heavy-metal, pesticide, and other noxious or undesirable residues. We do not have, or anticipate having in the foreseeable future, severe eutrophication problems arising in England and Wales, mainly because we have no lakes subject to heavy loadings with plant nutrients, and secondly because by and large we have not permitted the gross organic fouling of lakes, reservoirs and so on, which would greatly exacerbate the problem. In any case we have an effective and relatively cheap technology, for controlling the phosphate and nitrate (or ammonia) content of effluents, available for application where and when the need may arise.

The major problems lie first in establishing what types and maximum concentrations of residual chemicals are safely acceptable in public water supplies drawn from lowland, effluent-rich rivers, and second in establishing how to ensure that these maximum limits are not exceeded. A third major problem is to ensure that food fish drawn from rivers and the sea is not rendered acutely or chronically poisonous to man by water pollution. Many water scientists, including chemists, seem to stretch the magnitude of the public supply quality problem too far, by taking the line that a river water *might* be dangerously polluted, judged as a source of public water supply subject to long-term ingestion, because the contrary cannot be absolutely demonstrated, and since the cause of the danger is unknown controlling action cannot be applied. Therefore, goes the argument, water supplies should not be taken from lowland rivers but collected in upland reservoirs in protected gathering grounds and piped away for use. The conclusion of this argument has some force, but mainly for reasons other than those advanced.

There are important sources of water pollution other than effluent discharges. The rainfall run-off from paved surfaces in conurbations can be quite polluting as a result of the wash-off of minor leakages and spillages of contaminants at industrial premises and of atmospheric fall-out following a spell of dry weather. As our general technology progresses, so major accidental spills of toxic and noxious materials, many of which reach rivers or may add to groundwater pollution in regions of permeable superficial geology, increase in frequency. The run-off to rivers, and groundwater recharge, from agricultural land, appears to be containing increasing concentrations of nitrate as a result of intensifying agriculture, but the general low volume of rainfall over the last four years is at least partly responsible for this. In some areas recently, increased concentrations of nitrate in river waters and in abstracted groundwaters has caused very considerable interference with water supply operations.

Another danger of water pollution arises from the increasing production of solid and semi-solid toxic or noxious wastes for disposal. The dumping of such wastes in refuse tips, unless tightly and effectively controlled, can cause gross pollution of surface and groundwaters. Because of recent general realisation of the dangers of tipping toxic and noxious wastes on land, the current trend is towards provision of means for safe destruction or treatment of these wastes. By and large appropriate technologies for development and provision of these means of waste disposal can be devised without great difficulty, but the cost implications of this approach are very great.

For these reasons, it would be better, in water supply quality terms, not to take any public water supplies from lowland rivers; but there are many practical reasons, other than the loss of the economy of water re-use *via* rivers, militating against the application of this approach. The building of the very many upland reservoirs, of very many more miles of underground aqueducts, and the massive costs involved, are obvious and immense obstacles. Above all else, it would be folly to seek to do this for all public water supplies when about 99.5 per cent of such water, the proportion we do not drink out of the total used, does not need to be of such ideal quality.

An alternative approach, repeatedly canvassed, would be to provide dual, piped water supply systems, one of natural water for domestic drinking, and the second of second-class water but hygienic and of reasonable chemical quality for non-drinking domestic and industrial purposes. Another possibility, copied from the European mainland, would be to make arrangements for the supply of delectable natural water in bottles or cans for drinking, while mains supplies were kept wholesome if considerably less than natural. There are of course practical, economic, and social arguments against adoption of these alternatives.

The danger of rendering food fish poisonous to humans as a result of chemical contamination of rivers and seas has been clearly demonstrated by observation of such phenomena as bio-concentration and translocation of organo-chlorine pesticide residues through aquatic fauna, fish, and fish-eating birds, and the bio-methylation of residues of inorganic mercury and its concentration through aquatic food chains into fish. There are of course broader, but much less convincing arguments advanced in favour of reducing sea pollution, but these cannot be examined here.

In the case of river pollution, it is reasonable to associate the need to control this to prevent damage to river ecology and food fisheries with the desirability of achieving the maximum practicable re-use of water along the courses of rivers. It can be argued that if we make as certain as possible that our rivers are kept clean enough, and our waterworks purification processes adequate for, production of very good quality water supply, then we shall also go automatically a very long way to ensure full protection of the freshwater environment and to reduce sea pollution arising from river discharges to the sea. This implies that we can make a substantial contribution to our environmental survival at the apex of the ecosystem in the water pollution context, if we put ourselves right in the front line of exposure to pollution by using our lowland rivers as our main sources of public supply. Certainly to the layman the need to ensure direct protection of ourselves is infinitely more understandable, urgent, and deserving of financial backing than the need to protect ourselves indirectly at the selective end of an aquatic food chain.

However, in all these considerations, the basic question first to be answered is 'can we so control the quality of lowland river waters in the future that wholly satisfactory public water supplies can be increasingly produced from river sources, with concomitant increasing economy in water re-use?' The answer appears to be in the affirmative provided the principles of future national water policy, as recommended by the Water Resources Board, are augmented by increasingly stringent pollution controls, and by financial arrangements which, as far as practicable at any given time, minimise the growth of new demands on water resources and of water pollution.

The need for minimising the growth of new demands on water resources can be readily demonstrated. The lower the price of water supply, the more we tend to waste it, by unnecessarily polluting it and thereby not only degrading its re-use potential, but also causing river and sea pollution, by using it as a diluent for polluted process effluents, or by just letting good clean water run to waste. It is becoming increasingly urgent that these misuses of water be minimised. Bearing in mind that industry takes about four times as much water as is needed for domestic purposes, the need for water economy by industry is the most urgent, although domestic waste also is very important.

We need a new code of water pollution control appropriate to the increasing difficulties of improving or maintaining the quality of water supply rivers and of the water environment. This is to be provided by central government in a resuscitated Environmental Protection Bill.

We also need a drastic revision to be made in the national policy for charging for water use of all kinds, including the use of rivers for effluent disposal. The provisions of the Water Act 1973 and the above-mentioned Bill will permit such a revision to be made. The big problem here is to formulate the mechanism of a comprehensive, effective, fair, and socially acceptable charging scheme.

While it is not possible here to examine the various shortcomings of the present arrangements of charging for water use, it can be said that the main shortcoming is that the current price of water for use is much less than its overall environmental value. The essential needs of society for water supply and disposal can properly be regarded as part of this environmental value, but the luxury needs of society for water supply and disposal can be regarded as a wholly unnecessary debit on the environment, which should be heavily priced on that account. In short, essential domestic, industrial, and agricultural requirements for water supply and disposal should not be made any more expensive than the average costs involved in servicing these needs, while non-essential demands should be priced at a very high rate, perhaps equivalent to the marginal costs of securing new supplies and of disposal of the wastewaters arising. It is simple enough to decide that the quantity of essential domestic demand for water should be the current daily volume required by the bulk of households—i.e. o.60 m³/day per house of 4 occupants. It is however a much more complex matter to decide what is 'essential' in terms of industrial and agricultural use of water, but reasonable ways and means of doing this can be devised. Such an approach could avoid the considerable social opposition to metering of domestic supplies to households in the lower income groups—i.e. houses where use of significantly more than o.60 m³/day is unlikely, although metering in all other cases would be a necessity. There are of course other financial ways of inducing control of the rise of demand for water which cannot be examined here.

Future Organisation of Water Management

On 1st April 1974, a new organisation for water management came into operation as prescribed in the Water Act 1973. Prior to this there were some 1300 local authorities and joint boards concerned with sewage disposal, 200 public water supply authorities, and 29 river authorities concerned with water management. Clearly with so many autonomous bodies involved in water management it would have been extremely difficult to manage future requirements of large-scale water transfer, to enforce tighter pollution control, and to ensure optimum use of water for all purposes. Accordingly, proposals were put in hand just over two years ago, to place virtually all aspects of water management under the control of 10 Regional Water Authorities for the whole of England and Wales.

One very important factor involved in the formation of these new bodies is a new realisation of the key role to be played by science and scientists in water management, in which chemistry and biochemistry will be of central importance. The new authorities will not only be responsible for sewage purification and disposal, and hence be potential polluters of water, but also they will be responsible for pollution prevention. They will not only be involved in securing the best application of chemical process technology in sewage treatment and waterworks purification systems, but also in the detailed science of environmental conservation in the broadest sense. The proper discharge of these vital, and theoretically conflicting, roles by the new authorities demands that their scientific services should be such as to achieve a higher order of effectiveness and efficiency.

In addition to the formation of these new authorities, national research in water science has also been reorganised, as from 1st April 1974, in the formation of the Water Research Centre, comprising the former Water Research Association, the Water Pollution Research Laboratory, and the research arm of the Water Resources Board.

Overall therefore, the intrinsic relevance of chemical and other scientific research, technology, and management in meeting the water needs of society, has now been practically recognised; and it is mainly on the applications of chemistry by chemists, chemical engineers, and biologists that the proper conservation of water in the future depends.

Staying Alive

The war against hunger is truly mankind's war of liberation.

John F. Kennedy

Pallida Mors aequo pulsat pede pauperum tabernas
Regumque turris.

Horace

Protection and Preservation of Food Supplies

by R. F. Crampton; British Industrial Biological Research Association, Woodmansterne Road, Carshalton, Surrey

THE massive problems of world food shortage, the increasing number of individuals who are underfed or malnourished, and the increase in world population have been recognised and quantitated in approximate terms. The solution of these problems is becoming urgent, assuming that it is desirable to avoid the collapse of society in its present form. In the long term (though the time available is less than 50 years) practical measures of population control are essential. In the immediate future there are a number of ways to ameliorate the present situation. These include an increased rate of food production, and the so-called 'green revolution' is evidence of the efforts being made to achieve this. A second approach is to improve distribution of food which, in the opinion of some, is the major factor producing present-day malnutrition. The difficulties of achieving this seem to be economic and political rather than scientific. A third method is to reduce the amount of food which is wasted. Various estimates have been made of between a fifth and a third of all food produced being rendered unfit for consumption due to spoilage. This is surprising in view of the numerous methods available for the preservation of food.

Food preservation can be achieved by chemical and physical methods. The chemical methods depend upon the reactivity of specific compounds to combat spoilage due to infestation, bacterial contamination and autolytic changes which occur spontaneously in many foodstuffs. The physical methods produce changes in the physical state of the food which render it less amenable to spoilage.

A possible reason why chemical food preservatives are not used more extensively is the fear that harmful effects may be produced. The basis of this concern has stemmed from toxicological investigations, and the

Editor's Note: Unfortunately, the full text of Dr. Crampton's lecture is not available for inclusion in this book.

interpretation of data derived from them. Examples of such preservatives are sorbic acid, diethyl pyrocarbonate, sulphur dioxide, ethylene oxide, and nitrites. From such considerations two important questions arise. Firstly, does the desire to avoid adverse effects of chemicals to man lead to an over-conservative restriction on the use of preservatives? Secondly, is the philosophy of complete safety the correct basis for decision making, or should the concept of a risk–benefit ratio be adopted? The latter question is likely to be answered differently, according to whether the answer is given by an adequately fed or a starving individual. If this is true, the efforts which are being made to unify international regulations affecting the use of food preservatives may be ill-founded. This also applies to other means of protecting food supplies, and recent decisions of the western world on the future use of some pesticides may be quoted. The extent to which idealistic policies are contributing to, and inhibiting the solution of, a vast human problem needs to be reassessed.

Novel Sources of Proteins

by B. M. Lainé; Assistant General Manager, BP Proteins Ltd., London

Introduction

OVER the last 10 years a great deal has been said and written about the magnitude of the present and future world food shortage. This has often been received with scepticism, especially in developed countries. But, unfortunately for the sceptics, the problem is on our doorstep. We remember the uproar generated by President Nixon's decision, 9 months ago, to put an embargo on all exports of soya because of a shortage in the United States. The immediate impact of the embargo was a rapid rise in the price of pig and poultry meat, in this country and in Europe, because those animals are fed with a diet containing 15 to 20% soya meal, nearly all imported from the United States.

President Nixon's decision was primarily due to a bad harvest but, nevertheless, it underlined a basic problem: the world-wide increasing imbalance between an expanding population on the one hand, and available food resources on the other.

This paper considers the nature of the world food shortage particularly in terms of protein deficiency. It goes on to consider possible means of counteracting the shortages from conventional (meat, seed, fish) sources and from novel Single Cell Protein (SCP) production.

The nature of these novel processes is discussed and some of the costs involved in SCP production from different substrates are compared.

The World Food Shortage

The deficit is particularly acute for the basic food elements. It is not so severe for energy-providing cereals where a tremendous effort is being made to raise yields—the so-called 'Green Revolution'—but is acute for proteins, the essential component for organism development and tissue regeneration.

The protein shortage is developing rapidly, particularly in the underdeveloped countries of Asia, Africa, and South America. Its impact is becoming evident on the standard of health and working capacity of the poorest groups. If the shortage becomes worse, it will make these countries heavily dependent for continued existence upon the western world.

This situation is now recognised at the international level. United Nations experts agree that to cope with the magnitude of the world-wide problem, novel approaches will have to be considered seriously.

A few figures will help to put the problem into perspective. The world's population is increasing presently at an annual rate of 1.8% but this rate of increase is unevenly distributed, with 2.6% in Asia, South America, and Africa and only 1.1% in the USA, USSR, Europe, and Australasia. By 1985 we shall probably number a thousand million more. By the year 2000 present numbers will have doubled.

In many regions, people are relying for their diet on cereals which are poor in protein content (10% wt.) and of an unsatisfactory quality because of a balance of component amino-acids inadequate for human needs. For example a daily consumption of 500 g of cereals will provide 50 g of protein, of which the body will only use 25 g efficiently. To have full value, these proteins must be supplemented by others of different amino-acid composition, mainly of animal origin, e.g., meat, eggs and milk. It has been calculated that an average additional 25 g of such protein would be required to provide the daily 75 g sufficient for the average man.

Whilst in the western world we consume more than we require of good proteins (above 40 g/day) the average daily intake in many other countries is only around 10 g/day. FAO has calculated that to boost it to 16 g in 1980 and 22 g in 2000 an additional 15 Mt and 35 Mt will be required respectively by those dates (4% increase in availability per annum).

How can this be achieved? More cattle grazing could be achieved and would provide more meat, but this may be at the expense of cereal production since land is limited. The greatest effort is being put into pig and poultry breeding where a better protein yield per unit of feed can be obtained. This is so because these animals, like humans, are of the monogastric type and require for a fast growth a balanced diet of cereals and other rich protein sources such as fish meal or soya meal. Here again we find a dramatic competition for the use of available food sources. The FAO does not forecast more than a 2% increase per year in meat production.

More fish consumption might be a solution. But, during the last few years, the catch has been stable at the rate of 50 Mt/year of which 20 Mt are transformed to give 4 Mt fish meal. There are many doubts that it could be economically increased, except by fish farming in some specific cases.

With the help of FAO, Peru has become the major fish meal producer. The annual fish catch there has soared from 84,000 tons in 1948 to an almost unbelievable ten million tons in 1970, only 20 years later. But world fish resources are limited. Many experts think that already there are indications of overfishing. This is one of the reasons given for the reduction in the Peruvian catch to a mere 5 Mt two years ago and for the cancellation of fishing last year. These actions would give the anchovies a chance to grow bigger. We understand now why in December, 1967 the General Assembly of the United Nations Organisation sponsored a programme of work on the research and development of novel sources of proteins as being one of the few practical, efficient solutions of the problem we have considered so far.

Two main avenues of effort are open. Development of: oil seed protein; single cell protein (yeasts or bacteria similar to the ones we are using in BP).

Can oil seed production (peanut, soya, or cotton, for instance) be dramatically increased? The presence of toxic components limits their use for human food without further processing. The present practice is to crush the seeds, extract the vegetable oil, and treat the residue to get a protein meal. Such a meal is mainly used in animal feed. Additional treatment might make it more palatable and a valuable human food. But, to avoid an excessive vegetable oil production, the effort would have to concentrate on soya which gives 20 g of oil and 80 g of protein meal per 100 g of seed. Unfortunately, here again the possibilities are limited because of the shortage of adequate land, so that the experts do not expect a tremendous increase over the 20 Mt now produced mainly in the USA. It is interesting to note that an increase in production of 10 Mt would require the mobilisation of an additional 20 million acres of crop land.

Single Cell Proteins—a Possible Solution

Single cell protein, a neologism now widely used to cover protein derived from micro-organisms, may make a valuable contribution to the solution of this problem.

The use of micro-organisms for the transformation and production of foods is as old as the discovery of fermentation. Brewing has been known to man for at least 6,000 years, probably following Noah's experiments, and the use of brewer's yeast in baking for almost as long. The production of food yeast by fermentation was carried out in Germany as far back as World War I. Since then a small industry has developed, producing fodder yeast from molasses (mainly in Cuba) from the waste products of the food industry (such as whey or starch), and from carbohydrate waste in the paper industry.

But the main limitation to any large development has always been the quantity of waste available at any given point; it has always been too small to make fodder yeast production economic in its own right.

The advance came when Champagnat suggested to BP, in 1959, that hydrocarbons should be used in place of carbohydrates as substrate for yeast growth.

Various attractive features were associated with the use of hydrocarbons for the production of micro-organisms. First, hydrocarbons were then a cheap substrate—the n-paraffins which the micro-organisms assimilate often being an undesirable component in some crude oil distillate fractions.

Secondly, they could be made available in very large quantities, where protein was required, because of the ease of transportation.

Thirdly, micro-organisms are extremely efficient protein producers: they can double their number in a few hours; or, to use a more striking comparison half a ton of ox will produce about half a pound of protein per day, whereas half a ton of yeast will produce some $2\frac{1}{2}$ tons of protein in the same period.

Obviously, some of these features are worth reconsidering in the light of recent events, but this will be referred to later.

Processes for SCP Production

Let us now consider the processes that we in BP, and others in our wake, have developed.

In the baker's yeast industry an aqueous solution of carbohydrates (sugar) and minerals is fed into a reactor where yeast is present. Under suitable conditions of aeration, the yeast oxidises and assimilates the sugar and multiplies. When all the sugar is consumed, the additional quantity of yeast produced is harvested, normally by centrifugation.

Basically, in our process we have replaced sugar by normal paraffinic hydrocarbons and simple batch production with continuous production. Two approaches were possible. We could either feed middle distillates from crude oil to the reactor and let the yeast assimilate the paraffinic molecules or extract the n-paraffins by an absorption process beforehand and feed only these to the fermenter.

In the latter case, if the fermentation is correctly conducted, all the paraffins are consumed and the harvested yeast is ready for use after drying. In the former case, as the middle distillate is only partly paraffinic, the non-consumed hydrocarbons have to be separated from the yeast biomass produced.

Both processes need a purification stage, but in one case this is done beforehand and leads to a simple fermentation/harvesting process; in the other case purification comes after fermentation and leads to a more complex harvesting purification process.

The route using middle distillate has been developed at Lavéra, in the South of France, where the original breakthrough by Champagnat was made. In Scotland, at Grangemouth, the route using n-paraffins of high purity produced by the BP normal paraffins process has been developed. A very similar approach to this latter process is being followed by Japan and USSR.

In this country two variations of this type of process are being explored for the production of animal foods, by Shell and by ICI. Shell is developing a process using methane as substrate, but it would appear that they are still at the pilot plant stage and they look on it as a second generation process. ICI are undoubtedly more advanced with their process, based on methanol derived from natural gas. They have recently commissioned a 1,000 t/year pilot plant at Billingham.

SCP Substrates

There are many possible substrates for SCP production. We have already mentioned pure n-paraffins, the paraffinic components of gas oil, methane, or products derived therefrom—methanol, ethanol, acetic acid—have been considered and carbon dioxide could also be used for the growth of algae or plant tissue.

The selection of substrate will depend upon many factors, economics being one of the most important.

An even less conventional substrate is now receiving attention—manure. In 1972 approximately one billion tons of solid organic waste was generated in the US. Of this waste, nearly 200 million tons came from animal manure and this is posing great pollution problems. Feed lots handling up to 100,000 head of cattle are becoming common. The manure is stacked in huge heaps; a single one of these accumulations is roughly equivalent to the municipal sewage disposal problem for a city of 1 million people. It has been proposed to convert this manure into protein by using it, after preconditioning, as a substrate for bacteria—millions of tons of bacterial protein could be produced.

Present Status

At this point in time the products derived from both BP's processes are destined for animal feeding. This is the result of a deliberate policy decision taken years ago. BP have always been aware of the tremendous economic, sociological and, in some areas, the religious problems based on taboos that the introduction of such a novel source of food would raise. It was considered that the magnitude of these problems was such as to make unrealistic any large scale undertaking to produce human food in the near future.

On the other hand, the animal feeding industry is a very large one, even by our western European standards. Compounded animal feed production in 1972 was:

> 10.9 Mt UK
> 9.6 Mt France
> 9   Mt Holland
> 10 Mt Germany
> 4   Mt Italy

a total of 53 Mt for the EEC. In all these countries production is increasing rapidly, except in the UK and Holland, where such feed production started earlier and consequently stabilised earlier. Production is estimated to increase to 60—65 Mt by 1975.

Because of its protein content, 60—70% wt. according to the process, BP's yeast ranks as one of the richest protein ingredients for pig and poultry diets—similar to soya bean meal and fish meal. Consumption of fish meal and soya bean meal in the UK and Common Market has increased from 4.7 Mt in 1965 to 6.2 Mt in 1968 and is expected to increase to nearly 11 Mt in 1975. Because of fish meal's production limitations this is mainly in demand for soya.

There has been for years a distinct fear amongst animal feed compounders that pressure of demand will result in a very great pressure on price. Recent events mentioned earlier have confirmed this fear. This explains the compounders' interest in an industrial process producing a similar commodity which, because of its continuous nature, would ensure a much more regular supply of stable quality product with price, therefore, much less sensitive to demand in case of scarcity.

We commissioned a 4,000 t/year plant, using the pure normal paraffins route, at Grangemouth in 1971. We have also commissioned at Lavéra in 1972, a 16,000 t/year plant using the gas oil route. We are building a 100,000 t/year plant in Sardinia in partnership with ANIC, the Italian State chemical company, such a size of plant being in line, we believe, with the large potential market. Such a careful approach oriented first towards animal feeding is, in our opinion, the only realistic one to launch a novel source of protein; while implementing this philosophy we shall have time to prepare for entry to the human food market as required. But, in the interim period, a very large animal feed development will help to switch towards food some of the commodities presently used as feed, particularly fish, thus easing the protein shortage.

What are the Prospects?

At a recent UNIDO Expert Meeting on SCP held early in October, 1973, a comparison was made between the *cost* of various possible substrates. The report, prepared by Dr. Young of ICI and the present author, and endorsed by the group, can be summarised as in Table 1.

Table 1: Costs of Substrates

	Feedstock Cost $/t	Yield t SCP/t Substrate assimilable	Substrate cost $/t of SCP
Molasses (50% assimilable carbohydrates)	20—40	0.25	80—160
n-Paraffins	60—80	1	60—80
Gas oil	20—30	1	20—30
Methanol	40—60	0.5	80—120
Methane	5—20	0.6	8.5—32
Ethanol	60—120	0.6	100—200

To prepare this table we have used a range of figures; the low figure for molasses corresponds to a cost in producing countries, whereas the high one corresponds to a cost in importing countries, in particular Western Europe. The bracket of figures for methane reflects the possible difference of value at the crude oil well head in a producing country of the Middle East and in Western Europe where it is costed at fuel oil value.

Any economic assessment of SCP production must also take into account the oxygen requirement and heat evolved during fermentation. It is generally accepted that fermentation on methanol, paraffins and methane requires, respectively, 2, 2.5 and 5 times more oxygen per unit of cell than on molasses, the heat evolutions from the fermentations being roughly in the same ratios. Assuming that the power required for oxygen transfer to the cells is in the same ratio, the Table 2 showing energy costs per tonne of SCP product can be derived. For this purpose we have used the minimum and maximum figures published in terms of kWh per pound of yeast produced from

Table 2: Energy Costs

	Ratio of O₂ Requirement	Energy Consumption Costs $/t
Molasses	1	5—10
n-Paraffins	2.5	12—24
Gas oil	2.5	12—24
Methanol	2	6—20
Methane	5	25—50
Ethanol	2	6—20

molasses and assumed a cost of 1 cent/kWh. However, it must be noted that the efficiency of oxygen utilisation varies from process to process.

In order to make a comparative estimate of the impact of heat evolution on fermentation costs (Table 3) we have assumed that cooling water at 18—20°C is available (typical N.W. Europe sea-board location), that fermentation with yeast takes place at 30°C, whilst for methanol and methane, bacteria are used and grow at 40°C, hence increasing the heat transfer efficiency by a factor of 2.

Table 3: Cooling Costs in SCP Production

	Fermentation Temperature °C	Heat Evolution Ratio	Relative ratio of coolant Circulation	Cost $/t
Molasses	30	1	1	5
n-Paraffins	30	2.5	2.5	12
Gas oil	30	2.5	2.5	12
Methanol	40	2	1	5
Methane	40	5	2.5	12
Ethanol	40	2	1	5

If we add up the costs given in Tables 1—3, taking account of the fact that if gas oil is used, the costs must be multiplied by 1.2 because of the loss of lipidic material suffered during the final stage of purification by solvent extraction, we obtain the extreme costs shown in Table 4.

Of course, on top of all this must be added harvesting costs, chemicals, manpower, maintenance costs, depreciation and profit.

The group endorsed the conclusions which had emerged from the discussion at an earlier conference on SCP held at the Massachusetts Institute of Technology in May, 1973, namely: 'On the basis of the recent trend in price

Table 4: Total Cost/
Tonne of SCP at
Fermenter outlet

	$/t
Molasses	90—175
n-Paraffins	84—116
Gas oil	54—91
Methanol	91—145
Methane	45.5—92
Ethanol	111—225

of fish and oil seed meal there is every reason to be most optimistic about the future economic feasibility of SCP.'

Recent events in the Middle East have considerably obscured the horizon. In the author's opinion the problem is not associated with the availability of hydrocarbons. When we consider the demand for products like ours we are thinking in terms of millions of tons, and this is negligible in comparison with the current production of oil. It seems certain that, if the economics are right, hydrocarbons for protein production will find their place in the pattern of use of hydrocarbons. The pattern for the future reveals less use of hydrocarbons as a primary source of energy and much more use as a valuable source of carbon.

It is, at this point in time, difficult to forecast for how long or by how much, oil prices will continue to increase. It is also difficult to forecast the impact of this on fish meal or oil seed meal prices. It must not be forgotten that although seeds are produced by nature, their production requires fertilisers; their harvesting requires energy and a lot of machinery. Various experts foresee that, under the direct pressure of the increase in oil prices, under the indirectly generated inflation pressure, and because of the latent shortage, the increase in the price of these animal feed commodities will continue to rise sharply.

We in BP believe that the statement made earlier on the economic feasibility of SCP protein production from hydrocarbons will continue to hold true and in conclusion we feel we have opened the way for an attractive new field of operations, which might also contribute to the solution of one of the most dramatic problems facing the world.

Promotion and Regulation of Plant Growth

by David L. Gerwitz; Research Manager, Agricultural Products, Monsanto Europe S.A., Brussels, Belgium

Introduction

IT is appropriate that this session, which includes a discussion of plant growth regulation, is part of a session entitled 'Staying Alive'. Much of the applied, early work in this area was done in Britain and the United States during World War 2. In fact, Imperial Chemical Industries Ltd., represented here by Dr. F. A. Robinson, played a vital role at their Jealott's Hill Research Station. Surely, the research done during the war years was directed towards one objective—Staying Alive.

The approach I have chosen for this discussion on 'Promotion and Regulation of Plant Growth' includes three parts: the challenge to the chemical industry, a general view of what plant growth regulators are, and, thirdly, what plant growth regulator research entails on a practical level.

In a highly industrialised society, there is a tendency to underestimate the complexities encountered in agriculture. For many, contact with agriculture is limited to scenic views from cars during a holiday trip, displays of produce in the supermarket, or the food itself as it emerges from a bottle, can, or package. There is a widespread attitude that there can be no real problem producing food since fresh fruit and vegetables, flour, sugar and animal products are available year round. Apples and other fruit are casually discarded if they show a blemish. So, for many consumers the critical issues involve the variety and price, not the presence or absence of something to eat.

However, in spite of this apparent abundance, food production has historically been a precarious undertaking and continues to be so in too many parts of the world.

Primitive man must have realised the truth of this statement as he scavenged and hunted. Biblical man recorded his plight and resorted to prayer in order to escape the hardship of a meagre harvest. Solomon, himself, prayed for deliverance from blasting, or mildew; locusts, or caterpillars. The

79

Romans also resorted to prayer and they had a special god, Robigus, who ruled over the dreaded rust diseases that could destroy the cereal crops. The Robigalia was in fact a time of prayer for Robigus to withhold the devastating diseases.

This may seem ancient history which has little relevance in modern times, but in the space of this hour devoted to a discussion of plants and crop production, over 2,300 people will starve to death and the year's total will approach 20 million people. This prediction was made at a recent Rockefeller Foundation meeting in New York where information was also provided which showed that the 5 year drought in Africa has caused the sands of the Sahara to move southward at rates of up to 30 miles per year. This drought has already forced as many as six million people to lose their animals and crops, and the land that nourished them, and to return to the uncertain existence of a nomadic life much the same as that of primitive man. To many living now, history is reality and the presence or absence of food is a daily problem.

For the most part, farmers in the northern temperate regions do not ponder over the possibility of complete crop failures. Their endeavour, however, is far from being free from uncertainty because they are involved in a business activity which is rapidly changing due to factors which reflect, among other things, increasing international demands for more food of higher quality, and more localised problems such as reductions in manpower, spiralling costs of goods, and the necessity to maximise profitability.

Agriculture has responded admirably to these challenges. In the United States, for example, the agriculture work force has declined by over 50% during the past 20 years. Farm numbers have reduced by a similar amount and yet in the same time American farmers have been able to increase production by more than 500%.

Yields of crops are a reflection of a myriad of factors including:

1. Quality of land
2. Favourable climate
3. Timely precipitation or the availability of irrigation
4. Adequate supplies of fertiliser
5. Yielding potentials of seeds and plants.
6. Control of weeds, insects and diseases
7. Availability of machinery
8. Adequate energy
9. Adequate financing
10. Good management practices

The chemical industry may take pride in its accomplishments in assisting the farmer to reach even greater yield levels through the production of fertilisers, herbicides, insecticides, fungicides, nematocides, animal feed additives, animal health products and preservatives.

With the array of chemicals already available as well as the new herbicides, product mixtures, and systemic pesticides now in varying stages of research and development, it would be easy to feel complacent. However, the need for greater production of safe, highly nutritious food continues to escalate and a constant challenge to improve production faces everyone who is even indirectly involved with the agricultural industry.

The dramatic increase in corn yields in the United States is shown in Figure 1. During the forty year period from 1930—1970, yields have increased almost 400%, and agricultural chemicals played a key role.

In the late 1930s to early 1940s, hybrid corn appeared. Hybridisation brought hybrid vigour—the superior growth and yielding potential of progeny resulting from two genetically diverse parents. But, the genetic potential of hybrid corn could not be realised without additional inputs. This is also true for the dwarf wheats and rice which are key elements in the green revolution. Large scale fertilisation, especially high nitrogen-containing substances, became common in the late 1940s. The impact of insecticides, fungicides, and herbicides was felt about 10 years later. The result achieved by countless technicians who put together the package of agronomic practices is a national average corn yield over four times greater than that of 40 years ago.

Similar success stories can be told for cotton, potatoes, tomatoes, and many other crops. In the 30 year period from 1940—1970, the United States yield of cotton has almost doubled and potato yields have almost trebled.

The success story of corn can also be used to present a challenge to the agricultural community since the rapid increase in corn yields may be credited to several technological achievements including hybridisation, fertilisation, and pesticides. Will there be another breakthrough or will the curve begin to reach a plateau as the application of technology reaches maximum implementation?

Figure 1 also shows the slow increase in soybean yields during the same period. What about this very slow increase in a major source of vegetable protein and oil? Must the agricultural industry wait until the plant breeder develops hybrids, or soybeans which respond like corn to nitrogen fertilisers, before this crop begins to reach optimal production? Pick another crop and the same type of questions can be asked. What can the chemical industry do?

To many, plant growth regulators represent the next major contribution towards increased yields. Reduced to the simplest form of definition, these are products which, when applied at non-herbicidal rates, exert a controlling influence over plant growth. Some have explained these compounds in terms of chemical genetics—an analogy which may be sound based on the physiological and morphological effects already demonstrated by natural and synthetic plant growth regulators.

The promise offered by this area of research is significant. A practical breakthrough in any of the major world crops may have tremendous humane, economic, and political consequences. This is our challenge of the 1970s and a vital element in the needs of society.

The area of plant growth regulators is not new. Over 200 years ago Duhamel du Monceau studied plant growth and concluded that sap produced in one part of a plant moved to other parts and exerted an influence on growth. Put another way, sap or other substances produced in the leaves controlled the growth of roots.

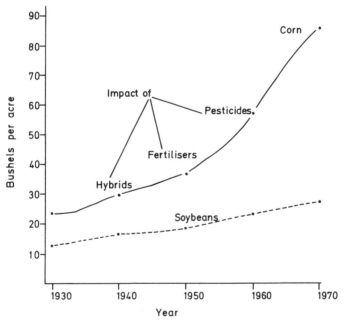

Figure 1: U.S. National Average Yield of Corn and Soybean, 1930—1970.

Darwin also contributed to this research by observing the influence of external forces, such as light, on the growth of plants. His work set into motion a vast array of studies which culminated some 50 years later in the discovery of IAA, indoleacetic acid.

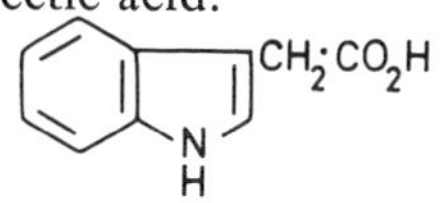

Indoleacetic acid (IAA)

In subsequent years it was shown that auxins, typified by IAA, influenced numerous aspects of plant growth including form, root growth, lateral bud growth, fruit development, callus formation, and the abscission of leaves and fruit.

During the war years, research in the United States and in Jealott's Hill led to a side benefit, namely the elucidation of selective suppression of plant growth and the development of the auxin or hormone herbicides typified by 2,4-D, 2,4-dichlorophenoxyacetic acid, and MCPA, 4-chloro-2-methyl-phenoxyacetic acid

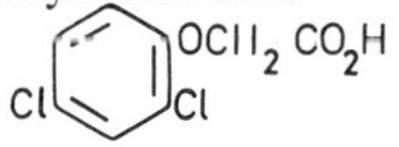

2,4-Dichlorophenoxyacetic acid   4-Chloro-2-methylphenoxyacetic acid
(2,4-D)   (MCPA)

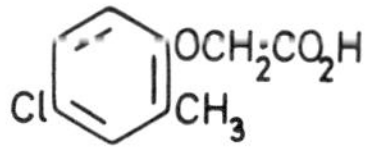

Once the practical utility of synthetic biologically active substances was recognised, research in plant growth regulation has followed two separate, yet closely related, pathways. Some have pursued the study of natural products in an effort to explain the intricate ways in which plant growth is controlled. Others have sought to utilise the concept to produce new methods through which food production could be enhanced.

In the more fundamental approach, attention has been focussed largely on the natural growth regulators or hormones which include the auxins, gibberellins, cytokinens, ethylene, and the inhibitors, one of which is abscisic acid.

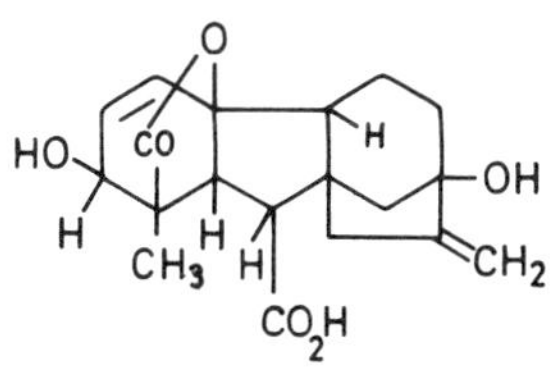

Gibberellic acid (GA₃)

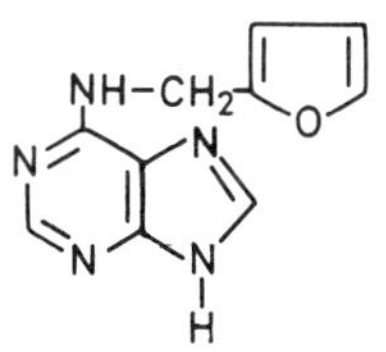

6-Furfurylamino purine (kinetin)

$H_2C=CH_2$

Ethylene

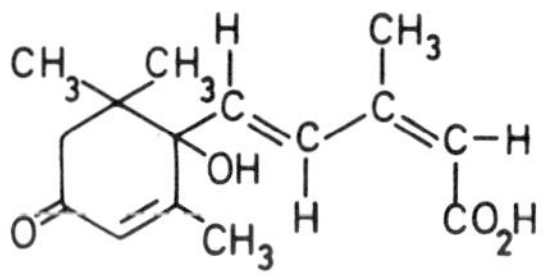

Abscisic acid (ABA)

The gibberellins were discovered because they played a key role in the expression of a plant disease. The Japanese had long struggled against the *bakanae* disease of rice which is caused by the fungus *Gibberella fujikuroi*. Plants infected by this fungus often grow as much as 50% taller than healthy plants and the literal translation of *bakanae* is 'foolish seedling'. The growth-promoting substance produced by the fungus was isolated and named gibberellin as early as 1935. Following World War 2 simultaneous studies in the United States and at ICI in England led to the chemical identification and naming of gibberellic acid, which is also designated as GA3.

Today more than 37 compounds of this class have been identified and shown to be common to an extremely wide variety of plant species. As a group, the gibberellins are especially active in increasing stem elongation and have found commercial value in the production of table grapes where the compound stimulates the production of larger and more elongated berries.

The first cytokinen, kinetin, was discovered due to the mysterious fact that coconut milk or yeast extracts stimulated the growth of plant callus tissue cultures. This type of regulator influences cell division, breaks seed dormancy, and causes a mobilisation of nutrients to areas treated with the product. 6-Benzylaminopurine has commercial utility in prolonging the shelf life of leafy vegetables.

Ethylene is now becoming accepted as a natural growth regulator. The mechanism of action is still not known but the application of the gas, or ethylene generating compounds such as ethephon, (2-chloroethyl)-phosphonic acid, has shown that this substance influences diverse phenomena such as the release of dormant buds in potato tubers, ripening of fruit, degreening of citrus, enhancement of sap production such as the flow of latex from rubber trees, the abscission of plant organs, and stimulation of flowering in cucurbits and pineapple.

Natural inhibitors are common in plants and highly diverse in structure. Their presence in plants should not be unexpected when one considers that nature has developed many feed-back systems in which the overproduction of a compound leads to the generation of a counteracting compound or system to keep the metabolic controls working in an orderly way.

Inhibitors such as abscisic acid have been shown to influence bud dormancy and disease resistance, and may provide an ecologically competitive advantage, as illustrated by juglone. This substance is produced in the roots and bark of the walnut tree and may inhibit the growth of neighbouring plants.

Research in the area of naturally occuring growth substances is beset by numerous problems including the extremely low levels of the compounds in plants, their transient nature, natural occurrence in a bound form, and the fact that nature seldom has allowed a control mechanism to exist independently but has integrated the action of the various hormones. One of the greatest challenges in this area is the elucidation of the mechanisms by which these substances interact.

In much the same way that the chemical industry provided herbicides, without first defining their mechanism of action, products to control plant growth are also evolving. The number of compounds already available precludes a discussion based on chemical groupings. A more interesting approach to the practical aspects of growth regulation is based on the situation in which a given crop is selected and the agronomist, plant-breeder, or farmer responds to the question of what should be changed to increase the productivity or profitability of the crop.

To illustrate this approach I have selected corn, *Zea mays*, because it is a major world crop and one which offers a significant humane and commercial opportunity for growth regulators. Table 1 presents an abbreviated list of the ways in which growth regulators could make a major impact on corn yield, profitability, or both. A casual review of the agronomic literature would

Table 1: Plant growth regulator opportunities in *Zea mays*

1. Seed production
   A. Male sterility
   B. Increased seed per plant
2. Growth characters
   A. Enhanced seed germination
   B. Longer/shorter maturation period
3. Environmental stress
   A. Drought resistance
   B. Disease resistance
4. Composition
   A. Increased protein level
   B. Increased sugar level (vegetable types)
5. Productivity
   A. More ears per plant
   B. More plants per acre
6. Harvestibility
   A. Reduced lodging
   B. Shorter dry-down period.

greatly expand this list and would illustrate comparable opportunities for other major crops.

Male sterility induced by chemicals could replace the cytoplasmic approach now utilised in creating seed plants devoid of pollen and thereby incapable of selfing. The time involved in generating suitable lines would be greatly reduced and certain limitations inherent with cytoplasmic sterility would be eliminated. The ability to modify sex expression is practical in the case of cucumbers where the use of ethephon results in exclusive production of pistillate or female flowers. This process also increases the seed yield per plant and provides a means of concentrating maturity for one-pass harvesting.

Enhanced seed germination could increase the rate of seedling development and could provide both an escape mechanism to avoid seedling rots and a means to obtaining better stands. In other species such as grape and cherry, GA_3 can break the rest period and induce germination. Stimulation of germination by GA_3 has also been demonstrated in peas and beans.

Modification of the maturation period for corn as well as many other crops can lead to better control over the harvesting period. The advantages could be one-pass harvesting, greater efficiency, and reduced losses. SADH (succinic acid 2,2-dimethylhydrazide) and ethephon have shown desirable effects on a range of crops including cherry and tomato. NAA, naphthaleneacetic acid, is a stop-drop for Bartlett pears and thereby extends the harvest period.

The potential benefit to be derived from enhanced drought and disease resistance is obvious. Major commercial success on field crops has not been realised but film forming agents such as waxes have proved valuable in preserving fruit during storage and shipment. They are also utilised to prolong the display life of house plants and to increase the winter hardiness of ornamentals. CCC(2-chloroethyltrimethylammonium chloride), ethephon, and Phosfon-D (2,4-dichlorobenzyltributylphosphonium chloride) have all been reported to enhance disease resistance in crops such as cotton, cereals, and potatoes.

Corn is a carbohydrate crop but, in a world facing numerous crises, including a protein shortage, the ability to modify the protein level in corn such as the changes in lysine content *via* the opaque lines, could mean higher values for both human and animal nutrition. Sub-herbicidal applications of triazine compounds have modified the protein levels in cereals. The potential to make sweet corn sweeter can be realised from the fact that sugar cane treated with glyphosine (*NN*-diphosphonomethylglycine) has higher levels of sucrose than untreated sugar cane.

Yields may be increased if a plant will yield more fruit. This has been shown to be possible in snapbean, soybean, highbush blueberry and cranberry with compounds such as 4-chlorophenoxyacetic acid, TIBA (2,3,5-tri-iodobenzoic acid), and NAA.

Increased populations and reduced lodging might be achieved from a reduction in plant size. MH (1,2-dihydropyridazine-3,6-dione), CCC, SADH, Phosphon-D, TIBA, and a range of other experimental products have been shown to be effective on lawn grass, shade trees, apple, snapbean, azalea, chrysanthemum, wheat, and tomato.

A faster dry-down period could facilitate an earlier harvest in the case of corn. Mechanical spindlepicking of cotton requires that either leaf moisture is reduced or that leaves are 'abscissed'. Numerous compounds are effective for this purpose including ammonia, ammonium nitrate, sodium chlorate, or paraquat (1,1-dimethyl-4,4-bipyridinium).

This exercise was presented to illustrate several points:

1. Yields of major crops such as corn can be increased and growth regulators may provide the mechanism to accomplish this.
2. Sufficient experience is already available to establish that growth regulators can be found to help the farmer improve his efficiency and gain better control over his operation.
3. In spite of all the illustrations given above of the effects of plant growth regulators, there are still major challenges which have not been met.

The pathway to greater productivity *via* growth regulators appears to be straightforward. Those having experience in the usual screening approaches by which industry has discovered and developed most of the pesticides, may expect that the approach to growth regulators would be similar. It may be. However, the experience of most as indicated by the negative results obtained in field tests, suggests that this endeavour will be more complex.

The usual pesticide screen involves the exposure of a test organism to a myriad of compounds. The identification of an active compound is relatively simple since the insects die, leaves are free from disease, or desirable species of plants grow while the weed seeds fail to germinate.

Herbicide tests may be designed to detect compounds which will dwarf plants and may therefore have utility as growth retardants. Similar simple tests measuring seed germination, inhibition of bud development, *etc.* can also be used to provide visual screens.

But, unlike pesticide field tests which are often short term, and small plot studies evaluated visually, many aspects of plant growth regulator studies defy this approach.

A good example is a product to modify the composition of a cereal. The effect in this case is not visually apparent, and the plant part to be analysed is only available after a minimum of several months of growth during which environmental, agronomic, and genetic factors introduce variability.

A successful effort in plant growth regulation must consider numerous factors including the following:

1. Responses may not be visual and may require multiple analytical measurements to follow the influence of a product.

2. Trials have to be designed to provide sufficient size, location numbers, *etc.* to overcome high levels of variability common to yield studies.

3. Rate and frequency of application must be evaluated and controlled since opposite effects may be obtained as the dose changes. IAA is a classic example of this effect, since stimulatory responses change to inhibitory responses as the level of the auxin increases.

4. Timing must be closely studied as a plant may move from a receptive to non-receptive stage within a short time span. Fruit thinning has been called an art because of this variability in plant–chemical interaction.

5. Non-uniformity in plant development under field conditions may cause some plants or even parts of the same plant to vary in development. Irregular maturation of olives may result in only partial effectiveness for a chemical designed to loosen the fruit.

6. Genus, species, or varietal differences may limit the value of a product. An example is the mixture of grass and broadleaf plants found in most turfs. A product used to reduce mowing of turf may do an effective job in limiting the growth of certain varieties of grass, only to have the desirable species eliminated because selective suppression by a dwarfing agent has swung the competitive advantage to the weed species.

7. Ignorance of the time at which significant morphological events take place, the interaction of one growth character with various external stimuli, or the type of physiological response required, may reduce growth regulator research to a 'spray and pray' programme. For this reason, it is necessary to have research in growth regulators proceeding along many lines.

The complexity of the research is obvious, but should not be overestimated. The plant breeder has been faced with comparable obstacles and yet has developed significantly better varieties. For the chemical industry, however, there may have to be changes in approach such as modification of the composition of the research groups to include a greater diversity in expertise. Co-operation between industry and the academic community must be increased to bring the available knowledge to bear on the problems, and management may have to be prepared to invest greater amounts of resources over longer periods of time.

Agriculture must continue its work to improve the yields of other crops just as corn has been improved in the past 30 years. The chemical industry has played a vital role as shown by the contributions already made in fertilisers, herbicides, insecticides, and fungicides and whether plant growth regulators will play as significant a role remains to be seen. However, we cannot become complacent.

The reality of starvation is too acute and the issue facing all of us is obvious —'Staying Alive'.

Staying Alive—and Healthy

by F. A. Robinson; President, Royal Institute of Chemistry

THERE are areas of the world where 'staying alive' is still the main preoccupation of the populace. One such area is the Wollo Province of Ethiopia where a serious famine occured in November 1973. But even those who survived the famine are likely to die at a relatively early age from one or other of the diseases endemic in the area—malaria, typhus, tuberculosis, cholera, smallpox, or typhoid. Life in this 'natural' environment is pretty tenuous, and man is competing for living-space, not so much with other large mammals, as with microscopic creatures, such as bacteria, spirochetes, and protozoa, that are far more lethal.

Disease in the Middle Ages

There is a great contrast in the expectation of life today in primitive parts of Africa and that in Western Europe today, but not a very great difference from the expectation of life in 17th or even 18th Century Europe. The most famous epidemic disease to affect Europe was the Black Death, more correctly described as the Great Pestilence, which hit these islands in 1348 and destroyed one-third of the population of the towns. Its probable birthplace was the foothills of the Himalayas where it spread across Asia and the Roman Empire of Justinian in 540 AD to disappear for the next 700 years. After its reappearance in Europe in 1348, further outbreaks occurred intermittently in various parts of the Continent during the next 300 years. There was a serious outbreak in England in 1603, causing the deaths of 30,000 people, two others in 1630 and 1636 and the final outburst of 1665 known as the Great Plague. The Great Fire of London in the following year probably helped to eliminate the pestilence from England—though it continued on the Continent for many years—by clearing away the congested filthy hovels in which people lived. For we now know that the bacterium *Pasteurella pestis* responsible for bubonic plague is conveyed to the human victim by a flea which is itself conveyed from house to house and from street to street by the rat.

Nowadays we think of malaria as a tropical disease, but it existed as an endemic disease in Europe until the beginning of this century and indeed it existed in England for some years after the first World War around the Wash and in parts of Surrey, Essex, and Kent, especially along the estuary of the Thames. It was the scourge of Rome throughout the Middle Ages, and caused the Popes to leave the Vatican Palace during the summer months for the healthier air of the hills. Columella, writing in the days of Nero, stated that 'a marsh always throws up noxious and poisonous steams during the heats, and breeds animals armed with mischievous stings which fly upon us in exceeding thick swarms . . . whereby hidden diseases are often contracted'— a startling anticipation of the discovery of the way in which the malaria parasite, *Plasmodium*, is distributed by the mosquito, and a concept that was forgotten for more than a millenium and a half.

Another disease that we also associate with the tropics is leprosy, but this too was endemic in England from the 11th Century onwards, having probably been introduced by the Normans, for the old name of the disease is Lepra Normannorum. It is not a particularly contagious disease nor a very deadly one, but it results in horrible disfigurement. Henry IV probably suffered from leprosy during the last years of his reign, but the disease appears to have died out by the end of the 15th Century.

Disease in the 18th and 19th Centuries

But as malaria and leprosy disappeared other diseases took their place. The dread scourge of the 18th Century was smallpox, for not only was it a killer, carrying off 1/13th of each generation, but those that survived were often badly disfigured. In the early 19th Century, cholera suddenly appeared in this country and other parts of Europe. It was first identified in Britain in Sunderland in 1831 and thence spread rapidly to Manchester, London, and other big cities. Its association with polluted sources of drinking water was deduced by Dr. James Kay in Manchester and Dr. John Snow in London, who checked further outbreaks in the City by removing the handle of the Broad Street pump. This quickly led to the recognition that cities required a clean, unpolluted source of drinking water and to the introduction of piped water supplies by local authorities. Cholera, except for sporadic outbreaks, has disappeared as a serious cause of death but typhoid, another water-borne infection, has continued to cause many deaths, including that of the Prince Consort in 1861, even up to the middle of this Century.

Nevertheless the health of the population had vastly improved by Victorian times and so had the expectation of life. Improved housing and hygiene, better food and a clean water-supply, improved medical and nursing skills all helped to remove the factors responsible for the rapid spread of contagious diseases during the Middle Ages. Smallpox was steadily coming under control with the wider use of vaccination. Nevertheless the conditions of the poor—especially the young—were appalling, as Charles Dickens pointed out in his emotive tales of Little Nell and Tiny Tim, Paul Dombey and Dora. The disease that caused most deaths in Victoria's reign was tuberculosis. This was responsible for one in five of all deaths, and about 60,000 people in Britain died from it every year. After 1900 it began to decline, partly because of more effective preventive measures and partly because of the increasing prosperity of the population.

Another Victorian scourge not much discussed at the time, because it was mixed up with sex and promiscuity, sin, guilt, and morality, was syphilis. The form we know today appears to have come to Europe from Central America with Christopher Columbus' returning sailors in 1493. This disease is of special interest in the present context, because it was the first to succumb to a synthetic chemical.

Syphilis is not a killer as is tuberculosis, and in 1900 the death rate in England and Wales was only about 150 per million. But the disease is highly emotive, since an infected man might infect his wife and through her their future children, and only 50 years ago 1200 infants died each year from congenital syphilis.

Chemistry takes a hand in the Fight against Disease

This then was the situation at the beginning of the present century—the Great Pestilence merely an unpleasant historical memory, malaria and leprosy no longer of importance in Western Europe, smallpox capable of being controlled by vaccination, whilst cholera and other water-borne diseases were becoming increasingly uncommon because of improved public health measures. The main causes of death were infections with various bacteria—diphtheria, pneumonia, puerperal fever, and of course tuberculosis—and viruses—measles, influenza, and smallpox. It was at this stage that chemistry first entered the lists in the battle against disease, although it was not until the late 1930's that chemists—who of course should not be confused with pharmacists, who play a quite different role in health matters—began to score dramatic successes. Unfortunately the public has little idea of what chemists have done to extend the expectation of life and make life more healthy and enjoyable The rest of my talk describes some of the contributions chemists have made to 'staying alive and healthy', in the hope of dispelling some of this ignorance.

Actually the first successes came not against bacteria but against a proto-zoon—the spirochete, the organism responsible for syphilis. Prior to 1900 this disease had been treated with ointments containing mercury, antimony, or arsenic, all of course highly toxic substances. The German chemist Paul Ehrlich working with the bacteriologist Robert Koch had discovered that some azo-dyes selectively stained micro-organisms and not animal tissues, and he conceived the idea of combining an azo-dye with arsenic or antimony to give a product that would be absorbed preferentially by micro-organisms and kill them without damaging the cells of the host. In 1905 he in fact succeeded in preparing a simple organic compound of arsenic, p-amino-phenylarsonic acid (I), which killed trypanosomes—protozoa that cause

(I) Atoxyl   (II) Arsphenamine

sleeping sickness—in the blood of experimental animals. This was intro-duced under the over-optimistic name, 'Atoxyl', but the substance was in fact quite toxic, and Ehrlich continued his synthetic work in the hope of finding a better chemotherapeutic agent. Two years later in 1907 he prepared arsphen-amine (II) known as '606' because it was the 606th compound he had pre-pared. Although not very effective against trypanosomes it was very effective against spirochetes and it was used in the treatment of syphilis under the name 'Salvarsan'. Together with its water-soluble methanesulphonate it remained the drug of choice in this disease for many years. It contains the $-As=As-$ bond closely analogous to the $-N=N-$ bond characteristic of the azo-dyes from which Ehrlich's researches started.

All kinds of consequences flowed from Ehrlich's pioneer work. In the first place, many other dyestuffs were synthesised by chemists and tested for antimicrobial activity, but unfortunately none of them had much effect on bacterial infections, although some, such as Trypan Red, Trypan Blue, and Bayer 205 had a beneficial effect in trypanosome infections. Other dyes, such as Methylene Blue, Crystal Violet, Malachite Green, and proflavine,

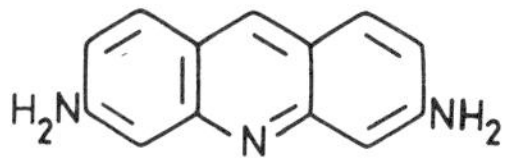

(III) Proflavine

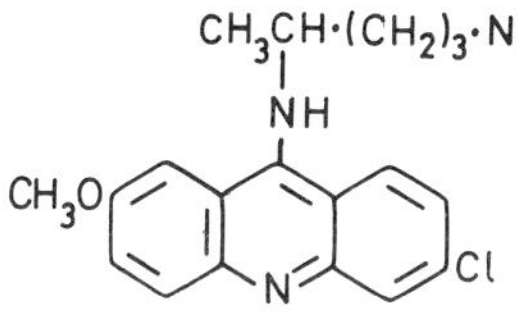

(IV) Mepacrine

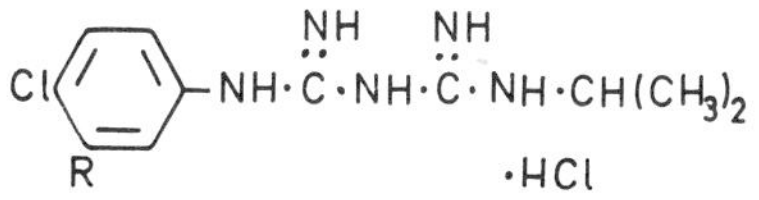

(Va) Proguanil (R=H)
(Vb) Chlorproguanil (R=Cl)

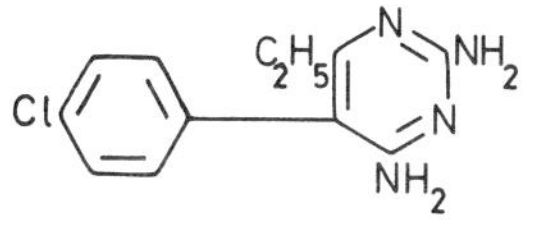

(VI) Pyrimethamine

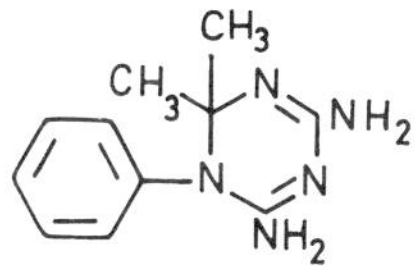

(VII) Metabolite of Proguanil

had antibacterial activity when applied to the skin but were toxic or ineffective when given by mouth or by injection.

The Fight against Malaria

At this point, developments took place in two different directions. One development started from Proflavine (III) and resulted in the synthesis of many new acridine derivatives. The introduction of dialkylaminoalkylamino-side-chains which were already known to be associated with antimalarial activity led to the synthesis by Mauss and Mietzsch in 1930 of the very effective antimalarial 'Atabrine' or mepacrine (IV) still in use today. Fortunately this compound was being produced in this country before the outbreak of the 1939—45 War, and its availability was an important factor in reducing casualties in the tropical war zones. Indeed so important were antimalarials regarded that a large programme of research was maintained throughout the War to find improvements over mepacrine, and a Committee was set up with Sir Robert Robinson as Chairman, to co-ordinate the work. The results were not published until after the War, but in 1946 a succession of papers appeared from Frank Rose and his colleagues of ICI describing the preparation of hundreds of new compounds in which every conceivable variant on the mepacrine theme had been made and tested. The most effective compounds that emerged from this work were proguanil hydrochloride ('Paludrine') (Va) and chlorproguanil ('Lapudrine') hydrochloride (VIb). These two compounds are more effective than mepacrine in preventing infection by the malaria parasite, but mepacrine is more effective for treatment. The reason for this is that the malaria parasite has a very complicated life-cycle with an asexual phase (schizonts) that undergoes a cycle of development in the red blood cells, a tissue form and a sexual phase (gametocytes) that undergoes a cycle of development in the mosquito. Substances affect the parasites at different stages in their life-cycle in different ways and, in addition, different species of *Plasmodium* vary in their susceptibility to the various antimalarial compounds.

This brief review of malaria chemotherapy would be incomplete without reference to George Hitchings' work in the USA, although this did not stem from either acridine derivatives or azo-dyestuffs, but from studies on the vitamin requirements of bacteria and other micro-organisms. This was one of my own special interests for some years, and we showed that many micro-organisms required for growth one or more members of the vitamin B complex, such as thiamine, nicotinic acid, pantothenic acid, or folic acid, and that their growth was inhibited by various antagonists that competed with the vitamins for coenzymes. Hitchings and his colleagues discovered that many pyrimidine derivatives were antagonists of folic acid and folinic acid for bacteria and *Plasmodia*, and as a result of their work pyrimethamine (VI) ('Daraprim'), was introduced in 1951 as the most effective of the pyrimidines for use as a suppressant in malaria. It is very potent and is effective when given once a week instead of daily as with proguanil. The diguanide chain of proguanil can be regarded as analogous to an opened-up form of the pyrimidine ring, suggesting a closer relationship between proguanil and pyrimethamine than would appear at first sight. That this is not just an attempt by the chemist to rationalise some fanciful idea of his own on structure/activity relationships, but a realistic concept, is supported by the isolation by Frank Rose and his colleagues in 1951 of a substituted phenyl-1,3,5-triazine

(VII) as a metabolite of proguanil, which proved to be a potent antimalarial in chicks, though unfortunately not in man.

The Fight against Bacterial Diseases

Now we must turn to the other development arising directly from Ehrlich's work, which was to have a much greater effect on Europe's ability to 'stay alive' than the discovery of the antimalarials. In 1935, Mietzsch and Klarer, reverting to Ehrlich's classical approach, prepared an azo-dye that contained a sulphonamide group. This was 2',4'-diaminoazobenzene-4-sulphonamide (VIII) which Domagk showed to be strongly bactericidal in mice especially

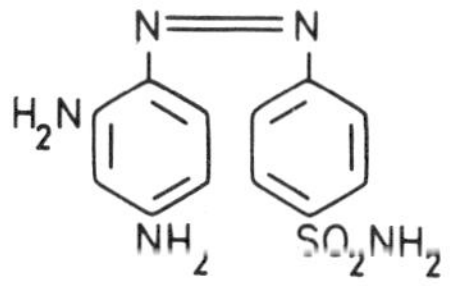

(VIII) Prontosil

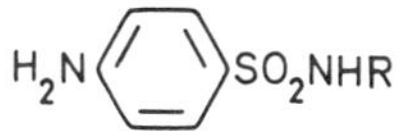

(IXa) R is H   Sulphanilamide

(IXb) R is 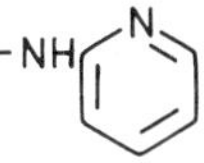   Sulphapyridine

(IXc) R is 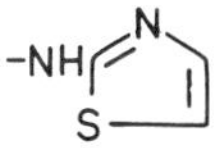   Sulphathiazole

(IXe) R is 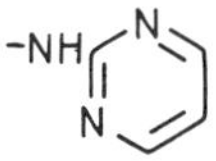   Sulphadiazine

(IXd) R is 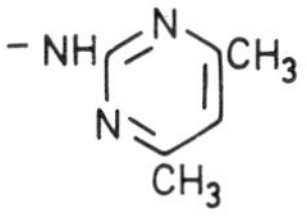   Sulphadimidine

(IXf) R is $-NH \cdot C \cdot NH_2$ with NH   Sulphaguanidine

against *Streptococci*, the bacteria responsible for puerperal fever, rheumatic fever, tonsilitis, and some of the pneumonias. It was effective by mouth, was not very toxic and was marketed under the name 'Prontosil'. It was suggested by Tréfouel, Tréfouel, Nitti, and Bovet that 'Prontosil' which had no effect *in vitro* was reduced in the body with cleavage of the double bond to give sulphanilamide (IXa); that this was in fact the effective antibacterial agent was confirmed experimentally by Fuller. Sulphanilamide had been made by Gelmo in 1908 but until 1936 no one suspected that it had any medical application. Shortly afterwards it was marketed and its success stimulated the synthesis of hundreds of related sulphonamides. The results were spectacular. In 1939 Ewins and Phillips at May and Baker synthesised M & B 693 or sulphapyridine (IXb), now superseded but at that time hailed as a miracle drug, and soon afterwards they synthesised sulphathiazole (IXc), which is still used in the treatment of staphylococcal infections. This was followed by sulphadimidine (IXd), the drug of choice in pneumococcal and streptococcal infections; sulphadiazine (IXe), the best drug in meningeal infections; sulphaguanidine (IXf), which is poorly absorbed from the gut and was used for a time in the treatment of bacillary dysentery; and phthalyl- and succinyl-sulphathiazole which have now replaced it. These are just some of the sulphonamides that were made and tested for this purpose—those in fact that have stood the test of time.

Many of the sulphonamides were tested in tuberculosis. But the organism that causes this disease, a *Mycobacterium*, is very different from *Streptococci* and *Pneumococci* that respond so dramatically to sulphonamides. It is a very slow-growing organism protected by a waxy coating, and of the sulphonamides tested only sulphathiazole and sulphathiadiazole had any useful degree of activity. However a sulphone, 'Dapsone' (X) showed some promise although later investigation showed that it was too toxic for clinical use. But its di-dextrose sulphonate was much less toxic and this was used clinically for some time under the name 'Promin'. Attempts to improve on 'Promin' by introducing other substituents into the ring or replacing one or both rings by heterocyclic nuclei were for the time being unavailing.

In the meantime in 1945, Lehmann had tested some 50 derivatives of benzoic acid against tuberculosis and found that one of them, *p*-aminosalicylic acid (XI) (PAS) had a high degree of activity. PAS had been made by Seidel and Bittner in 1901 and so, like sulphanilamide, it was a known chemical with hitherto unsuspected biological properties of great value.

There followed an intensive search for other compounds with antitubercular activity—a search more vigorous and widespread than any ever undertaken for a single disease except for cancer which in any event is a complex of diseases. As I have mentioned, sulphathiadiazole (XII) had some effect on tuberculosis and if the ring is opened one obtains the skeleton of a

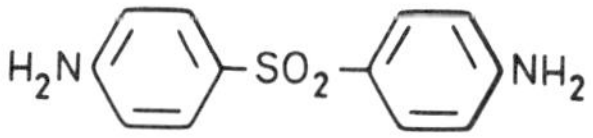

$$H_2N-\langle\rangle-SO_2-\langle\rangle-NH_2$$

(X) "Dapsone"

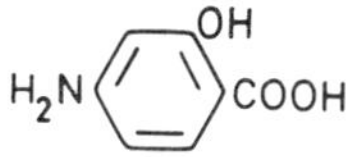

$$H_2N-\langle\rangle\substack{-OH\\-COOH}$$

(XI) PAS

$$H_2N-\langle\rangle-SO_2\cdot NH-\langle\rangle$$

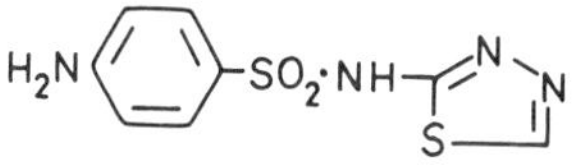

(XII) Sulphathiadiazole

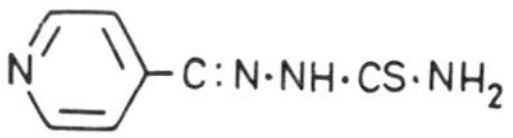

$$N\langle\rangle-C:N\cdot NH\cdot CS\cdot NH_2$$

(XIII) Isonicotinic aldehyde thiosemicarbazone

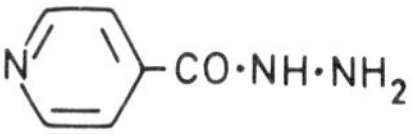

$$N\langle\rangle-CO\cdot NH\cdot NH_2$$

(XIV) Isoniazid (INH)

thiosemicarbazone. This was the next breakthrough, for several thiosemi-carbazones were found to be active and out of hundreds made and tested those of acetylaminobenzaldehyde (1948) and nicotinic and isonicotinic aldehydes (XIII) (1951), possessed outstanding activity. Then in 1952 it was found that isonicotinic acid hydrazide (INH) (Isoniazid: XIV) used as an intermediate in the preparation of the thiosemicarbazone of isonicotinic aldehyde, was as effective and less toxic than the latter. Once more a known compound—for it had been made in 1912—had been shown to have an important medical application.

Antibiotics continue the Struggle

But now to complete the story of tuberculosis and indeed the story of 'staying alive', I must describe another important development that originated from quite a different approach from Ehrlich's. So far the work I have described is that in which the chemist has taken the initiative and in which his synthetic skills have been paramount. Even so one must not forget that without the assistance of his biological colleagues—pharmacologists, bacteriologists, and toxicologists—the properties of the substances they made must have remained undiscovered. Admittedly in the early days testing was a relatively simple matter but later, as with malaria for example, the biological evaluation was often very sophisticated and success depended on a close partnership between chemist and biologist.

In 1929 Fleming, a medical bacteriologist, found that a mould spore that had fallen on to a Petri dish containing a *Staphylococcus* culture left unprotected on his bench at St. Mary's Hospital had germinated and that the area surrounding the mould mycelium was free from bacteria. He argued that the mould, *Penicillium notatum*, had produced an antibacterial metabolite to which he gave the name penicillin. He tried unsuccessfully to isolate the substance but he was able to show that the crude culture fluid on which the mould had been grown cured infected mice. It was left to Florey and Chain and their colleagues to isolate penicillin, and they reported in 1940 that their crude preparation cured mice infected with *Streptococcus pyogenes*, *Staphylococcus aureus*, and *Clostridium septique*. A year later they reported the first successful clinical trials, and even when supplies ran out they were able to continue treatment by extracting penicillin from the patient's urine and injecting it back into the patient. By this time, with the war situation growing grimmer, it was obvious that penicillin had enormous potential, and as one of the handful of chemists who had at that time handled moulds, I was greatly attracted to the idea of helping to develop a commercial process for making penicillin. The Directors of Glaxo Laboratories agreed and my group worked alongside Florey and his colleagues and other industrial groups to provide material for the Armed Forces. This we succeeded in doing with a great deal of ingenuity and improvisation, since everything one needed except apparently water, air, and glass seemed to be in short supply! We all pooled our resources and know-how and I became Secretary of the Penicillin Producers Research Committee, which was set up to facilitate communication amongst those working on the problem. Penicillin was in fact made available in time for the North African campaign. Because of our difficulties in the UK Florey had approached several manufacturers in the United States for help, and they used their know-how on deep fermentation to build much bigger production units than we had. It was then agreed to have Anglo-American co-operation in a study of the chemistry of penicillin and attempts to synthesise it. I became a member of the Committee for Penicillin Synthesis of which Sir Robert Robinson was Chairman and we used to exchange monthly reports with our American colleagues. The work was regarded as of such importance that it was made subject to the Official Secrets Act and the reports went across the Atlantic in both directions in the diplomatic bags of the two countries, and the work was ultimately published, not in the scientific literature, but as a separate monograph. Imagine my astonishment and chagrin when years later I met a business colleague in Germany who not only knew about these reports but disclosed that as a member of the German Intelligence Service during the War a copy of our reports had found their way on to his desk in Mexico a month after leaving London!

Actually the work proved far more complicated and difficult than any of us could have realised. We identified a whole family of penicillins and, although the most important member, benzylpenicillin (XV), was eventually synthesised in the laboratory, it never became possible to do this economically on the large scale. In due course however—actually nearly 20 years later—the chemist came into his own again with the preparation by Peter Doyle and his colleagues at Beecham Laboratories of a large number of semisynthetic penicillins (XV) made by acylating penicillanic acid. As a result the medical applications of the penicillins were considerably extended and it is now possible to control infections caused by bacteria that have acquired resistance to the older forms of penicillin; this had become a very serious problem.

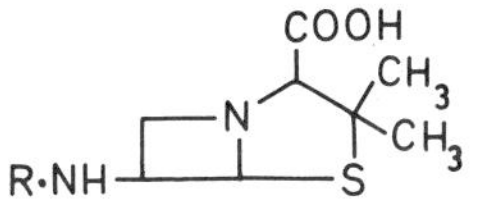

(XV) Penicillin

Benzyl penicillin $R = C_6H_5 \cdot CO \cdot$

Phenoxymethyl penicillin $R = C_6H_5 \cdot O \cdot CH_2 \cdot CO \cdot$

Phenethicillin ("Broxil") $R = C_6H_5 \cdot O \cdot CH \cdot CO \cdot$
CH_3

Methicillin ("Celbenin") $R =$

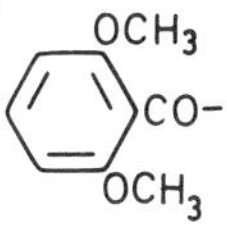

Ampicillin ("Penbritin") $R = C_6H_5 \cdot CH \cdot CO \cdot$
NH_2

Penicillin however was only the first of a large range of new antimicrobial agents which we now call collectively antibiotics. The discovery that a common mould could produce such an important medicine as penicillin started a search for other micro-organisms that might produce similar substances, and in this field the microbiologists took the initiative usually with the biochemist to help him and with the chemist coming in at the end to determine the structure of the antibiotic and if possible synthesise it and make derivatives. Although there are now over 20 antibiotics in regular use as medicines and they too have contributed to our ability to 'stay alive', I shall only mention one other—streptomycin—because this is necessary to complete the story of tuberculosis. Streptomycin was discovered by Waksman in the Institute of Microbiology at Rutgers University in 1944. He isolated it from the culture fluid in which an actinomycete, *Streptomyces griseus*, had been grown and he found that it was a remarkably potent anti-tubercular agent. Unfortunately, *Mycobacteria* soon developed resistance to streptomycin and this limited its use for a time. However, as I have already mentioned, both PAS and INH are also very effective anti-tubercular agents, and by using a combination of two of these compounds, or better, all three together, it is now possible to cure all forms of tuberculosis, although it is a somewhat slow business. The death rate in this country from tuberculosis was 1130 per million in 1920, 460 per million in 1940 and 60 in 1960, and sanatoria are things of the past thanks to chemists and their biological and medical colleagues.

As a kind of epilogue, let me add that leprosy which is caused by an organism closely related to *Mycobacterium tuberculosis*, has profited from the researches on tuberculosis therapy and 'Dapsone' (X), which was rejected for use in the latter is now the standard treatment for leprosy which can be cured given sufficient time, 3 to 9 months for tuberculoid leprosy and up to 5 years in lepromatous leprosy.

Of course medicinal chemists during all these years did not confine themselves to the discovery of substances to control the killer diseases for at the same time they succeeded in preparing substances that relieve pain, cure anaemias, induce sleep, prevent allergies and assist in alleviating many other conditions that produce a lot of misery and hardship and loss of earning power. In addition to contributing to 'staying alive' therefore the chemist has also contributed a great deal to 'staying healthy'. In fact the first discovery based on organic chemistry goes back to 1887 when acetanilide (XVI) was made and shown to be a useful agent for reducing the temperature in cases of fever. One of its metabolites was identified as *p*-aminophenol and this was acetylated, giving the *N*-acetyl derivative, paracetamol (XVII) (1893). This was found to be a good analgesic or pain reliever, as was phenacetin (XVIII) made in 1899 by acetylating *p*-phenetidine. The latter was thought to be the better compound and it remained the standard ingredient of headache tablets until a few years ago when it was replaced by the older analgesic, paracetamol, which is now known to produce fewer side-effects. Aspirin

Analgesics and antipyretics

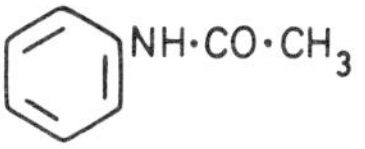

(XVI) Acetanilide

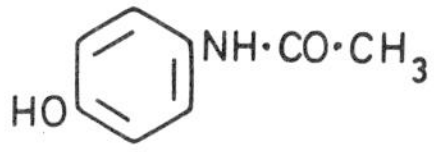

(XVII) Paracetamol

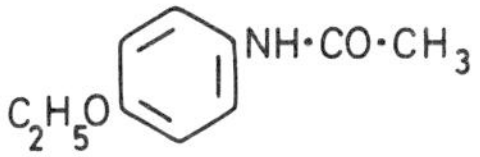

(XVIII) Phenacetin

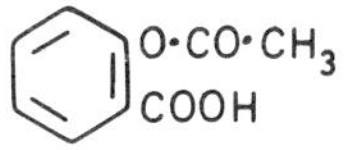

(XIX) Aspirin

(XIX), made by acetylating salicylic acid, was made by Kaufmann in 1909 and has been in regular use ever since. In spite of the fact that it produces gastric bleeding in some patients it is used, often in large doses, for the relief of that disabling condition, arthritis.

Time does not permit me to describe in any detail the steps that led to the decisions to synthesise particular molecules for specific purposes, but many new compounds were discovered because it became necessary to test every potential medicine for freedom from adverse side-effects. In testing certain sulphonamides on animals, for example, it was noted that some had an insulin-like action and lowered the blood sugar levels. This stimulated the

Hypoglycemics

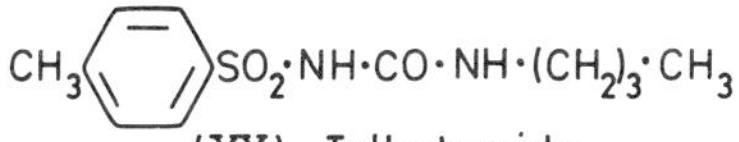

(XX) Tolbutamide

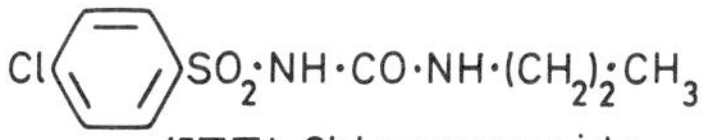

(XXI) Chlorpropamide

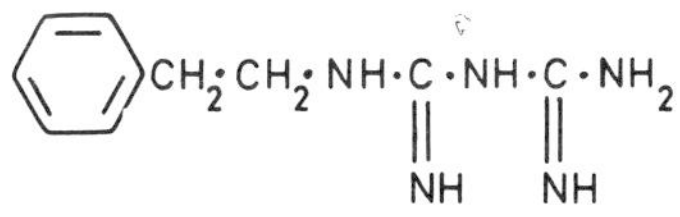

(XXII) Phenformin

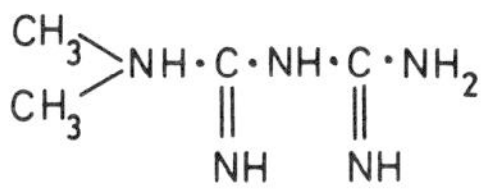

(XXIII) Metformin

preparation of other sulphonamides that might have a more potent insulin-like activity, and led to the discovery of tolbutamide (XX) (1956) and chlorpropamide (XXI) (1958). Then it was found that the sulphonamide group was not after all essential for activity and that the ureido-group could be replaced by a diguanide group, leading to the discovery of phenformin (XXII) (1960) and metformin (XXIII) (1962). Such are the imaginative leaps that sometimes lead to success! Other sulphonamides were found to stimulate the excretion of urine and this led to the discovery of new diuretics

Diuretics

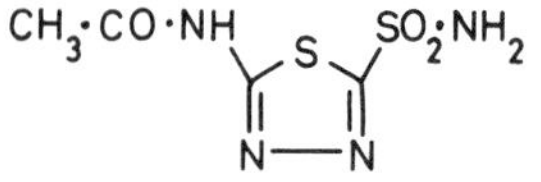

$CH_3 \cdot CO \cdot NH$ $SO_2 \cdot NH_2$

(XXIV) Acetazolamide

$H_2N \cdot SO_2$ Cl $H_2N \cdot SO_2$ F_3C $CH_2 \cdot C_6H_5$

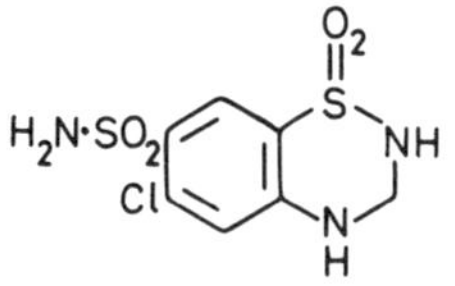

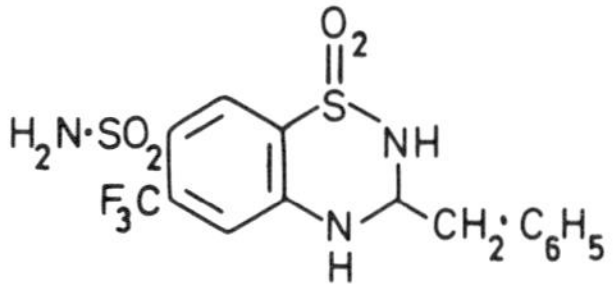

(XXV) Hydrochlorthiazide   (XXVI) Bendrofluazide

—acetazolamide (XXIV) (1950), chlorothiazide (1957), hydrochloro-thiazide (XXV) (1958), bendrofluazide (XXVI) (1959), and hydroflu-methazide, all useful in promoting the removal of fluid from water-logged tissue and thereby relieving the strain on an over-worked heart.

Other leads to new medicinal substances were provided by natural compounds with valuable biological properties. Their structures were not so much imitated as caricatured. Thus stilboestrol (XXVII) resembles—more or less—the steroidal oestrogens (XXVIII), some synthetic muscle relaxants

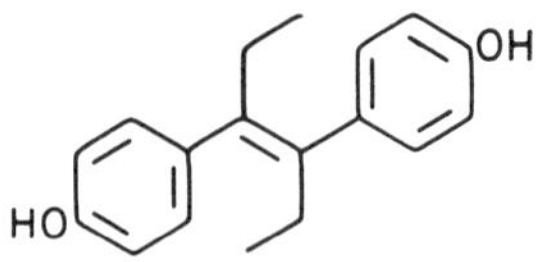

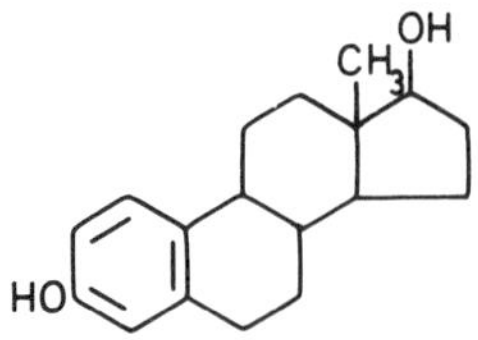

(XXVII) Stilboestrol   (XXVIII) Oestradiol

bear a remote resemblance to the South American arrow poison, tubo-curarine, whilst pethidine (XXIX), a synthetic analgesic, has a structure that

Analgesics

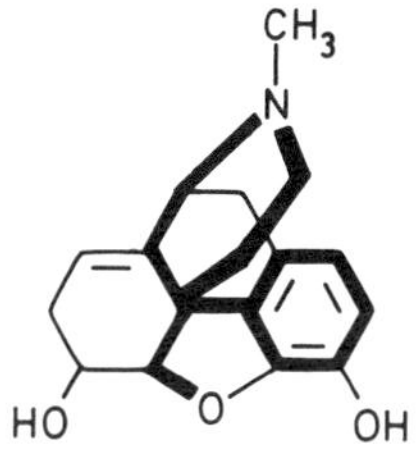

CH$_3$

(XXX)
HO   O   OH   Morphine

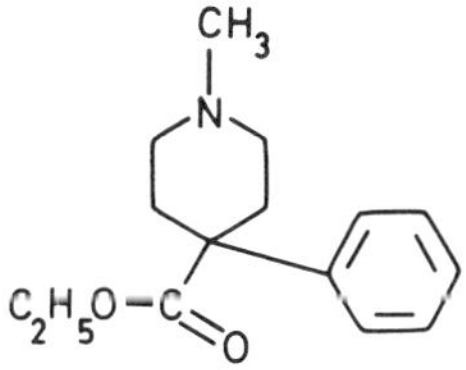

CH$_3$

C$_2$H$_5$O—C $\lessgtr$ O   (XXXI)
Pethidine

is an integral part of the morphine molecule (XXX), although the connection may not be obvious at first sight! Unfortunately the preparation of substances based on other parts of the morphine molecule failed to give anything of medicinal value, so perhaps pethidine was after all just a lucky accident! However it is a very potent analgesic resembling morphine, but unfortunately like morphine it is a drug of addiction and so must be reserved for the relief of intense pain. In fact there is a big gap between such mild analgesics as aspirin and paracetamol and the potent morphine type.

A Tragedy: and Legal Constraints

In 1957 it appeared that this gap might have been bridged with a non-addictive analgesic more potent than aspirin. This was α-phthalimido-glutarimide (XXXI) which was used for several years as a sedative under the name 'Thalidomide' before it was realised that it caused damage to the limb-buds of the foetus when taken by a woman in the first few weeks of pregnancy, so that the babies when born have no arms or legs or only stunted limbs with rudimentary fingers and toes. The association between 'Thalidomide' and these distressing foetal abnormalities was not at all obvious and was due to a persistent investigation by a young Australian physician, Dr. McBride, who published his conclusions in a paper to the *Lancet* on 16th December, 1962. This created a sensation. The drug was withdrawn from the market, laboratory

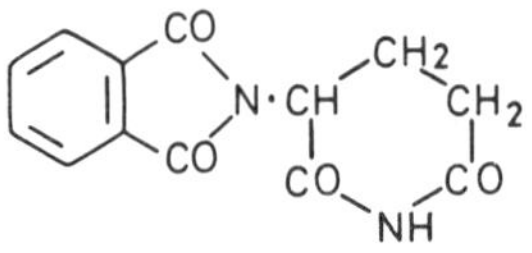

(XXXI) "Thalidomide"

investigations were put in hand to see if the effect could be reproduced in animals and, in due course, teratogenicity testing became part of the routine examination of every potential new medicine. In this country a Committee on the Safety of Drugs was set up to ensure that all potential medicines were thoroughly tested before being marketed. The control of new medicines was tightened still further by the Medicines Act 1968 which required that evidence of efficacy as well as safety should be submitted to the Committee before a medicine was released for sale.

Financial Constraints

This, and similar legislation in other countries, has made the task of the medicinal chemist very much harder than it was. In addition of course, the discovery of any new medicine makes it more difficult to discover a better one and, with the years, the areas where therapy is still inadequate have become fewer in number. Up to 1935, the screening of a few hundred compounds might yield two or three useful medicines, between 1935 and 1948 the chances of success fell to say 1 in 1,000 and by 1964 when the Committee on the Safety of Drugs was set up we used to reckon the chances at 1 in 3,000—4,000. Now, with the more stringent requirements of the Medicines Act, I would put the figure at nearer 1 in 10,000. This means that one must synthesise and test 10,000 new compounds to obtain one useful medicine. Of these, perhaps 9,000 will be rejected as being more or less inactive or too toxic and of the 1,000 survivors which are examined more thoroughly another 900 will be discarded as being no better than medicines in current use. Of the 100 that remain only ten (say) will survive the tests for chronic toxicity and teratology and half of these will fall down on metabolic studies and tests for cancer-promoting activity. Thus of the original 10,000 compounds only about five will be tested in human beings, first in volunteers and then in patients. The sole survivor will be put on the market and with luck will command sufficient sales to pay for the work on the 9,999 useless compounds!

The medicinal chemists' lot and that of his biological and medical colleagues are not in future going to be happy ones. It is impossible to guess for how long industry will be prepared to invest the large sums of money required to undertake this kind of research when the chances of success are so low. Indeed is society justified in expending so much skilled effort when, as I have tried to show, we now have the ability to control so many of the old killer diseases and provide so many aids to 'staying healthy'?

My own belief is that although the system of mass synthesis and random screening, known rather disparagingly as 'molecular roulette', has been successful in producing the wide range of excellent medicines now in regular use—and in spite of the length of this review, I have only mentioned a small number of them—it cannot continue indefinitely. In practice compounds are not selected just at random. Most medicinal chemists have considerable knowledge of the groups or side-chains that might be expected to enhance biological activity, and most of us have evolved hypotheses of our own to help us in our work. Often these hypotheses are found to be wrong, but because it is now customary to screen every compound for a wide range of biological activities, the unexpected occasionally happens and a substance turns out to have an application quite different from that for which it was designed! It is indeed a fact that most of the 467 new compounds introduced world-wide between 1958 and 1970 were discovered as the result of mass screening such as I have described, and that the most successful medicines were discovered by those companies that produced the largest numbers of new compounds and therefore spent the most money. Even so, the medicinal chemist would be much happier if he knew more about the chemistry of the human body, how the compounds he makes act on organs, tissues, cells, membranes, and the chemicals inside the cells, and so enable him to adopt a more scientific approach to his problems. I have said little about cancer in this talk, but most people believe that the attempts to find the proverbial cure for cancer have been a failure, in spite of the vast programme of testing that was mounted in the United States. More compounds have been tested as tumour inhibitors than for any other human disorder yet no single substance has been found that will cure cancer. However, a number of compounds have been found that will inhibit tumour growth for a limited time and by using these in succession or in a particular combination it is now possible to prolong the lives of cancer patients considerably and enable them to live almost normal lives.

The Need for a New Approach

Even more important perhaps is that we probably know more about the way

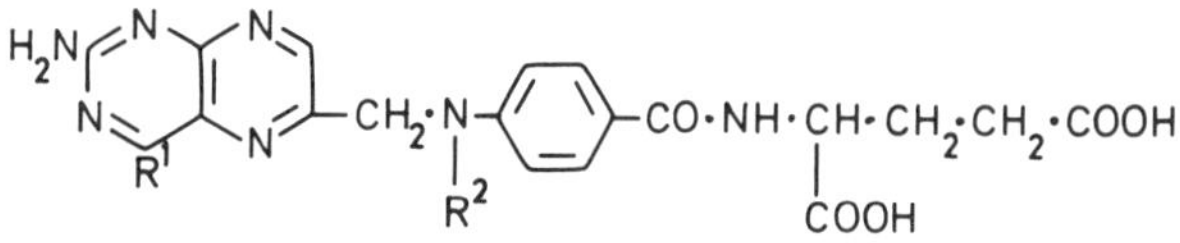

(XXXII) a Folic acid $R^1 = OH$; $R^2 = H$
        b Aminopterin $R^1 = NH_2$; $R^2 = H$
        c Amethopterin $R^1 = NH_2$; $R^2 = CH_3$

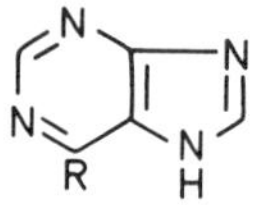

(XXXIII a) Mercaptopurine($R = \cdot SH$)
(XXXIII b) Adenine   ($R = \cdot NH_2$)

in which some of these anti-cancer compounds work than we do about the mode of action of many other medicinal substances. Thus aminopterin (XXXIIb) and amethopterin (XXXIIc) are folic acid (XXXIIa) antagonists that will produce complete remissions in 22% of children with acute leukaemia probably by inhibiting the dihydrofolate reductase in tumour tissue. Mercaptopurine (XXXIIIa) also produces remissions in children with acute leukaemia, but by a different mechanism, probably by competing with adenine (XXXIIIb) in nucleic acid biosynthesis. By combining amethopterin and mercaptopurine with other agents remissions in children can be increased to 90%, and the remissions can be maintained for at least a year by suitable adjustment of the dose of each component and proper supportive care. Tumours can of course be induced by chemical agents and by viruses and investigations in this field are leading to possible new methods for preventing the development of tumours. It is thought probable that some viruses may exist in a latent form which is activated by certain chemical agents and that the nucleic acid of the virus then becomes integrated with the DNA of the host cell, so modifying its behaviour that it becomes cancerous. This has suggested several lines of attack, for example the use of anti-viral agents, the development of immunological methods, selective inhibition of reverse transcriptase which occurs in both human leukemic cells and cancer-inducing viruses, and inhibition of the polymerase that produces DNA.

Nor is it only in cancer research that ideas for a more rational approach to the design of new medicines are emerging. The isolation from various sources of a group of substances known as prostaglandins which stimulate inflammatory reactions, induce fever, and act as 'local hormones' affecting restricted parts of the body, was followed by the demonstration that aspirin inhibits the release of at least two prostaglandins, thus indicating the way in which aspirin reduces fever. This concept of 'local hormones' may one day assist the medicinal chemist to design molecules able to react with specific receptors, and so produce the precise response needed. Another prostaglandin appears to initiate muscular contraction of the uterus and promote the onset of labour. If, as seems likely, premature labour—the most common single cause of infant mortality—is brought on by an alteration in the relative proportions of prostaglandin, oestradiol, and progesterone, then it is possible that a method of preventing premature labour may soon be made available.

Other developments that are showing promise have resulted from work on the structure of proteins, and it is now possible to synthesise peptide molecules comprising more than 100 amino-acid units. Already, results of practical value have been achieved with synthetic analogues of the adreno-corticotrophic hormone and recently comparatively small peptides that trigger the release of this and other hormones from the pituitary gland have been isolated, identified, and synthesised. A peptide has also been isolated from bee-venom and shown to have anti-rheumatic activity. All these peptides may in due course give useful leads to further advances in medicinal chemistry. Finally, four enzymes derived from natural sources—streptokinase, urokinase, 'Arvin', and 'Reptilase'—are currently showing promise in the treatment of arterial thrombosis.

The Fight must continue

In spite of the hundreds of medicinal substances now in regular use, we are still unable to provide adequate control over cancer, muscular dystrophy, arthritis, and migraine, for example. Kidney disease is still an intractable problem costing this country £80 million per annum in lost pay and, in spite of the existence of dozens of excellent tranquillisers, anti-depressants, and anti-anxiety compounds that have already improved the chances of recovery for many mentally ill patients, a satisfactory treatment for epilepsy, schizophrenia, and phobic disorders, still eludes us.

The skills of the medicinal chemist are therefore still needed, but I do not think we in this country can afford to continue to use the methods of the past. However the Government grant to the Medical Research Council is £27 million per annum and the total spent by Government on all medical research is £65.5 million per annum. A high proportion of this expenditure will be used, one hopes, to throw light on the physiological and biochemical basis of various disorders and disabilities. It would be of great benefit for this country if the results of these fundamental studies could be examined with a view to finding new leads for the design of more effective medicines, and any ideas were passed on to our pharmaceutical industry to provide an alternative to the expensive game of molecular roulette.

Enjoying Life

Human life is everywhere in a state in which much is to be endured, and little to be enjoyed.

Samuel Johnson

Mustard is no good without roast beef.

Chico Marx

Chemistry and Aging

by Charles G. Kormendy; Kali-Chemie Pharma, Hannover,
Germany (formerly of Bristol-Myers, U.S.A.)

In the broad sense of the word aging is a ubiquitous phenomenon which
affects not only the animate but also the inanimate world. Crystals, metal
alloys, colloids, and polymer fibres can also age.

While aging processes in physical systems of known composition can be
explained in precise chemical terms e.g., by the laws of thermodynamics,
decay of isotopes, changes in crystal structure, autocatalytic chain reactions,
etc., aging of living systems cannot, on account of their complexity, yet be
reduced to a single chemical event. Furthermore, because of phylogenetic
differences it is either inappropriate or dangerous to extrapolate to higher
species from experimental observations made in a lower form of life. More
bluntly, one might ask how far are we from translating animal data to man?
To paraphrase Dr. Alex Comfort we might have become excellent mouse
gerontologists although human aging is essentially the same old mystery.

On the positive side we ought to acknowledge the steadily increasing interest
in aging research particularly during the past 10 years which undoubtedly
will intensify in the future as society strives to improve the quality of life.

Biological aging is defined in most textbooks as progressive loss of
function, and increased incidence of disease and death of an organism with
the passage of time. The key element of this definition in the context of
'normal' aging is the progressive loss of function. It is expressed in the
organism's reduced capacity to adapt to the influence of intrinsic as well as
extrinsic factors. Thus, adaptability at a given stage of ontogenetic development reflects biological rather than chronological age.

For example, one such temporal expression of aging is manifested in the
induction of certain metabolic enzymes. When rats of different ages are
severely fasted, glucokinase, one of the glycolytic liver enzymes, becomes
completely depleted from the enzyme pool. Upon feeding the animals with
glucose, the activity of glucokinase reappears, after a brief delay, rises sharply,

and eventually reaches normal prefasted levels. While in young rats the inductive period is brief, in the old it persists for several hours. It was also observed that the time required to induce glucokinase correlates linearly with chronological age. It should be pointed out that old animals are fully capable of synthesising hepatic glucokinase to the same extent as the young, but it takes them much longer to achieve peak enzyme levels. A similar age-dependent lag characterises the induction of drug metabolising enzymes, such as the phenobarbital induced NADPH: cytochrome c reductase system. It is for this reason that drug therapy can be potentially more hazardous in elderly patients.

The capacity to respond to foreign antigens also changes with age. The antibody forming potential of mice sharply increases in the postnatal and growth period, reaches peak levels in adulthood, and then declines sharply during aging.

Certain adaptive deficiencies can be corrected experimentally. Notably, insulin administered to old rats abolishes the delay in glucokinase induction; several substances such as synthetic polynucleotides enhance the antibody response to foreign antigens; so-called 'catatoxic' agents reduce drug toxicity by accelerating the metabolic inactivation of drugs.

Lifespan, one of the most frequently used parameters in gerontological studies, reflects the overall adaptability of an organism to intrinsic and extrinsic factors. While the examples described above reflect changes in *vitality*, in lifespan the rate of *mortality* of a given population is measured against chronological age. For instance, adaptation to calorie-poor nutrition or lowered environmental temperature can profoundly alter lifespan.

In a series of well-known experiments McCay *et al.* kept colonies of rats on a calorie-restricted diet and compared their lifespan, growth, and disease patterns with those of controls on regular diet. Severe calorie restriction increased the lifespan of rats by more than 300 days. However, the dramatic increase is not the unique result of slowing down aging *per se*, but it is manifested in an overall retardation of maturation and growth.

Here one should also mention the recent ethnographic studies in Vilcabamba, Ecuador, where many of the inhabitants are centenarians. Their daily diet, consisting mainly of vegetables and fruits, amounts only to about 1700 calories. Of course, it is virtually impossible to prove whether the sparse diet is the dominant factor in their spectacular longevity or whether their genetic make-up predisposes them to an unusually long life.

Liu and Walford tested the influence of environmental temperature on the survival of the annual fish, *Cynolebias*. They found that lowering the water temperature by 5—6° C doubled the lifespan of this species. Furthermore, fishes kept at lower ambient temperature grew at a faster rate and reached a larger size than those at higher temperature, and the properties of collagen reflected a physiological retardation of aging. However, fishes are poikilo-thermic, and a suitable chemical agent has yet to be found which could induce prolonged, low-grade hypothermia in homotherms without loss of normal function. Marihuana constituents are currently being investigated in mice by Liu and Walford.

A Model of Aging *in Vitro*

Aging is not necessarily restricted to intact organisms. In tissue culture, Hayflick and Moorhead observed an aging-like phenomenon using human diploid fibroblasts. Cells were isolated from embryonic lung-tissue, cloned and maintained in an adequate nutrient medium. Under these conditions the cells divide until they cover the entire surface of the culturing vessel. At confluency, one-half of the cell layer was harvested and subcultured in another vessel. When the process was repeated serially, always in a ratio of 2:1 (doubling) the cells exhibited a distinct growth pattern, namely: adaptation or phase I, logarithmic growth or phase II, finally mitotic arrest and cell death or phase III. The total number of doublings was always limited to 50 ± 10, and the culture was free from infectious agents which could have caused cell death.

Hayflick and others reported that fibroblasts from adult donors undergo not only significantly fewer doublings than those from embryos, but the doubling capacity seems to decrease in proportion with the age of the donor. Also, fibroblasts taken from patients suffering from premature aging syndromes (*i.e.*, progeria and Werner's syndrome) barely survive in culture for more than two doublings. Thus, mitotic arrest and cell death might be caused either by cellular regulatory failure or activation of a mitotic inhibitor (*e.g.*, chalones). However, environmental limitations of the artificial milieu cannot be excluded as possible artifacts. More recently Holliday and Tarrant reported that errors indeed accumulate in enzyme proteins of late passage fibroblast cells, suggesting possible regulatory failure. Undoubtedly their findings will stimulate research along these lines.

One of the serious shortcomings of tissue culture systems is due to the fact that the cells are detached from the influence of neurohumoral regulation, and are maintained artificially. Consequently, extrapolations from tissue culture to organismal life might not necessarily be correct. Furthermore, only in-depth cytological and biochemical studies could establish whether the Hayflick tissue culture system represents a true model of aging.

Interference with Aging on the Functional and Structural Level

In contrast to normal dividing cells like the fibroblasts, most vital organs of the intact organism consist of postmitotic cells, which shortly after birth become morphologically and functionally differentiated and lose their proliferative capacity. Such cells compose the brain and nervous system, muscle tissue, liver, kidneys, *etc*. It is believed that these cells, because of their fixed state, are most vulnerable to aging. In other words, damage to postimitotic cells may result in permanent impairment or even cell death if self-repair cannot restore normal function.

The most frequently mentioned sources of age-dependent cellular damage are changes in the DNA–protein complex leading to functional modifications, or accumulation of errors along the critical path of protein biosynthesis, and structural deterioration by random chemical reactions, such as free radicals and cross-linkings.

Protein Synthesis

To date, we lack clear evidence that in an intact *vertebrate* species aging leads to mis-specification of vital protein components. Nor can we generalise that the codes carried in DNA or RNA for specific proteins are irreversibly altered because of aging.

Free Radicals

Excessive formation of free radicals and their interaction with cellular components is another suspected cause of aging.

Chemically, free radicals are highly reactive species which are generated by (*inter alia*) peroxidation of organic substrates. The peroxidative process is a chain reaction which consists of initiation, propagation, and termination steps, yielding inert end-products.

It is known that radiation can elicit free radical reactions in biological systems causing, for instance, alterations in the structure of DNA. However, eucaryotic cells are equipped with efficient enzymatic mechanisms to repair damaged DNA strands.

From the aging point of view, a more subtle occurrence of free radicals has been suggested. The primary site of free radical formation in cells appears to be in the mitochondria, which perform the oxidative respiratory function. Schematically, biochemical peroxidation has been rationalised in the following steps: (1) polyunsaturated lipids in the presence of biocatalysts such as cytochromes yield peroxide intermediates which are subsequently degraded to malondialdehyde, a highly reactive substance; (2) malondialdehyde either tautomerises to the less reactive β-hydroxyacraldehyde or it condenses with proteins, leading to conjugated Schiff-base adducts.

Interestingly the spectroscopic properties of such cross-linked protein derivatives resemble the so-called 'age pigment' or lipofuscin. In addition, the intermediary peroxides could also disrupt membrane structures, like the membranes of the nucleus, mitochondria, and lysosomes.

Based on the possible involvement of free radicals in aging, Harman initiated a series of lifespan studies using a variety of known antioxidants as free radical scavengers. Butylated hydroxytoluene (BHT), a commercially used food additive, at a daily dietary supplement of 0.50% increased the mean lifespan of mice by about 45%. Comfort evaluated the effect of ethoxyquin, which was originally developed as a rubber antioxidant. According to mortality data 0.5% of dietary ethoxyquin extended the survival of mice in both sexes by about 100 days.

Do BHT and ethoxyquin truly decelerate aging, or is it perhaps that both agents simply reduce food intake with the net result of slower growth and development as in McCay's calorie restriction experiments? The answer to this question is not clear, but BHT at oral doses equivalent to a dietary concentration of 0.25% which is one-half of the amount required to increase lifespan, produces appetite- and weight-loss in rats. In Comfort's study the ethoxyquin-treated animals weighed significantly less than untreated controls. Since both ethoxyquin and BHT are potent hepatic microsomal enzyme inducers, they might also accelerate such metabolic functions as, for example, the rate of lipolysis.

Conclusive answers to these questions ought to come from experiments evaluating the effect of antioxidants on biochemical and physiological parameters which are known to change during aging. In addition, it would be important to examine the effect of other antioxidants as well, particularly those having known physiological roles, such as vitamin E and glutathione.

A histological correlate of aging seems to be the progressive accumulation of 'age pigment' or lipofuscin within the cell. Direct as well as indirect evidence strongly suggests that lipofuscin is a depository of denatured cellular debris such as peroxidised lipids and membrane fragments and probably free-radical-mediated malondialdehyde–protein adducts. Although it is not proven that lipofuscin de facto interferes with normal cellular function, its preponderance during aging, particularly in the neurons of the brain and in the heart muscle, might have more than histological significance. For example, the percent volume occupied by lipofuscin substantially increases with age in certain regions of the brain of the mouse and rat. It should be pointed out that the hippocampus, which is supposedly the location of higher cognitive processes, shows the greatest pigment accumulation in both species. I might also add that Batten's disease which is characterised by massive lipofuscin deposition in the brain is accompanied by severe mental impairment.

Experimentally, it is possible to inhibit lipofuscin deposition with pharmacological agents such as meclofenoxate, kawain, and magnesium oratate.

Meclofenoxate was developed in France for the treatment of presenile and senile mental disorders and is currently available on the European market. Nandy and Bourne in the US, and subsequently German investigators, reported that meclofenoxate reduces brain lipofuscin content in old animals. The drug had to be given chronically for several weeks, with best results obtained after 10 weeks of treatment. Meclofenoxate increased also the mean as well as the maximum lifespan of mice. Kawain, a mild CNS stimulant, and magnesium orotate, a pyrimidine base metabolite, also prevent lipofuscin formation and restore learning behaviour in encephalopathic rats.

The real significance of these findings, however, needs to be evaluated on the basis of correlating biological dysfunction with lipofuscin accumulation.

If lipofuscin is the product of mitochrondrial degeneration as suggested, then its excessive accumulation may reflect a loss of mitochrondrial oxidative function which at a critical stage could lead to cellular hypoxia. Recently cerebral and cardiac hypoxias have received considerable attention in chronic mental and cardiovascular deficiencies, particularly as they occur frequently in elderly patients. For example, senile patients exposed to hyperbaric oxygen treatment showed a definite though temporary improvement in their mental performance.

Collagen

Collagen, an important structural protein, provides support and resilience to the skin, tendon, bone, cartilage, and connective tissues. Its aging is perhaps one of the best understood biochemical and physiological processes in gerontology. Since the underlying mechanism of collagen aging involves a type of intermolecular cross-linking, the phenomenon served as a convenient though somewhat tenuous argument for the so-called cross-linking theory of aging. Let me emphasise that collagen cross-linking is not a cause, but simply a consequence of aging. Nevertheless, the known principles present interesting possibilities not only in cosmetical applications, but also in the treatment of pulmonary, vascular, bone, and joint disorders prevalent in aging.

The collagen fibre has a triple helical structure arranged from three-alike polypeptide chains. Intramolecular crosslinks are formed within each of the three strands, conferring the proper strength and plasticity on the fibre. In contrast, intermolecular crosslinks tie the neighbouring chains together. The biochemical steps in the formation of intermolecular cross-linking are: (1) oxidation of lysyl residue to the aldehyde in one of the poly-peptide strands; (2) condensation of the aldehyde with the free amine of a lysyl group from another strand, yielding an intermolecular crosslink; (3) reduction of the Schiff-base to a more stable secondary amine. The mechanism involved in the last step has not yet been clarified, though most likely it is enzymatic.

The stability of collagen increases with the number of intermolecular crosslinks formed. Other properties, such as solubility and swelling capacity, decrease with greater stability. Thus, collagen stability (measured, for instance, by its tensile strength) is one of the best known indicators of physiological aging. However, according to lifespan studies, the overall rate of aging did not change when rats were chronically treated with β-aminopropio-nitrile, a lathyrogen and powerful inhibitor of collagen crosslinking.

Autoimmunity

I would like now to turn to the question of autoimmunity and aging. Walford proposed that aging might be regarded as an autoimmune condition involving a mild but chronic autorejection process. In other words, the organism progressively fails to distinguish its own cells from foreign cells. Although the hypothesis is not proven, disturbed immunosurveillance during aging is indicated by an increase in certain autoimmune factors, and increased renal lysozyme levels which normally rise after immunisation and trans-plantation. Also, removal of the spleen doubles the lifespan of mice while spleen cells from old animals injected into young ones shorten their survival time. It is conceivable that the spleen might harbour age-specific autoimmune factors. Amyloids which consist of globulins, collagen, and mucopolysaccharides are a suspected by-product of immune reactions and are frequently found in old animals and patients suffering from senile and presenile dementia.

On the assumption of an autoimmunological theory of aging, Walford evaluated the effect of azathioprine, an immunosuppressive drug, on the

120 CHARLES G. KORMENDY

lifespan of mice. While a daily dietary dose of 100 mg/kg increased the survival by 10 weeks at the point where 50% of the animals died, the maximum lifespan remained unchanged. Since azathioprine is a toxic drug, it is possible that liver toxicity obliterated the beneficial effects of immunosuppression. Similar results were obtained by cyclophosphamide, another toxic immunosuppressant.

Among the corticosteroidal immunosuppressive agents, prednisolone had a marked effect on the lifespan of a short-lived, highly inbred strain of mice. Bellamy found that prednisolone given in small amounts in the drinking water, from early life on, more than doubled the lifespan of the males. Although females responded also to prednisolone, generally they died earlier than the males. Since this mouse strain is suspected to have a moribund autoimmune disease, immunosuppression by prednisolone might account for the observed increase in lifespan.

As in the case of antioxidants, reduced food intake due to the prolonged administration of azathioprine, cyclophosphamide, and prednisolone cannot be excluded from consideration as a possible life-prolonging factor.

Hormones.

Historically, perhaps no other subject has received more attention from the gerontological and geriatric point of view than hormones. The literature, scientific or otherwise, contains countless experiments with glandular extracts and hormones ranging from naive and sometimes disastrous rejuvenation attempts to the medically more tolerable, palliative hormone replacement therapies. Unquestionably, the endocrine system which regulates and maintains the delicate homeostatic balance of the organism must *a priori* influence the overall aging process. However, the problem is not a simple one to study, since potential age-associated deficiencies may occur at any one point of a complex system such as from the central stimulus on to hormone production, release, delivery, and target activity. The problem is complicated by the lack of adequate biochemical techniques for accurate measurement of active and available hormone levels. Generally the sensitivity of tissues to hormones (*e.g.*, to insulin) declines with advancing age.

Body electrolytes, which are controlled by the posterior pituitary and adrenocorticoid hormones, tend to shift from the intracellular to the extracellular milieu with loss of potassium and retention of sodium. Friedman and his co-workers demonstrated that the rate of mortality decreases in rats when the electrolyte balance is restored with a crude posterior pituitary extract.

The indications are that oxytocin, one of the posterior pituitary hormones, might be the active principle.

Other hormones, such as pituitary growth hormone and thyroxine, failed to alter the lifespan of animals.

Conclusions

I hope that I conveyed to you in this brief review the current trends and thoughts which might apply to an amelioration or eventual control of aging. Undoubtedly the subject is diffuse, since by its inherent nature it encompasses many scientific disciplines in chemistry and biology. In this sense, experimental gerontology is in a fortunate position because it can draw new ideas and fresh approaches as they become available from other non-gerontologically oriented fields. For instance, a new line of gerontological investigation might evolve from recent discoveries made about the suspected role of latent viruses in the aetiology of cancer and also in certain dementia syndromes.

But a more imminent question is what should be done with the existing leads in terms of developing them into prospective practical payoffs. Lifespan studies are liable to criticism on conceptual grounds; namely, rate of mortality does not express the progressive loss of vitality which characterises normal aging. In other words, chronological and biological age do not necessarily coincide on each and every level of organismal function. If prolongation of lifespan by the methods which I have described represents an authentic slowing down of aging, then the effect should be measurable on a variety of levels of biological function. Such in-depth studies would require a long-term concerted commitment from scientists, governments, and private institutions to gerontological research before experiments in humans could be ethically as well as scientifically justified.

Would the effort be a worthwhile investment? I would like to answer this question by paraphrasing the eminent chemist and centenarian Professor Emmet Reid, author of the book *My First 100 years—An Interim Report*: 'The crowning achievement of research is adding more and better years to our lives.'

Bibliography

Bender, A. D., Kormendy, C. G., Powell, R.: Pharmacological control of aging. *Exp. Geront.*, 1970, 4, 97—129.
Comfort, A.: Longer life by 1970? *New Scientist*, 1969 (Dec. 11th) 549—550.
Comfort, A.: Basic research in gerontology. *Gerontologia*, 1970, 16, 48—64.

Comfort, A., Youhostsky-Gore, I., Pathmanathan, K.: Effect of ethoxyquin on the longevity of C3H mice. *Nature*, 1971, **229**, 254.

Davies, D.: A Shangri-la in Ecuador. *New Scientist*, 1973 (Feb. 1st) 236—238.

Hansan, M., Glees, P.: Genesis and possible dissolution of neuronal lipofuscin. *Gerontologia*, 1972, **18**, 217—236.

Hochschild, R.: Effect of dimethylaminoethyl *p*-chlorophenoxyacetate on the lifespan of male Swiss Webster albino mice. *Exp. Geront.*, 1973, **8**, 177—183.

Holliday, R., Tarrant, G. M.: Altered enzymes in aging human fibroblasts. *Nature*, 1972, **238**, 26—30.

Houck, J. C., Weil, R. L., Sharma, V. K.: Evidence for a fibroblast chalone. *Nature New Biol.*, 1972, **250**, 210—211.

Knappwost, H.: Das Altern in der unbelebten Welt; pp. 14—29 *in* 'Das Altern' (Lectures at the Joachim Jungius-Gesellschaft der Wissenschaften, Hamburg, 28—29th October, 1965). Vandenhoeck & Ruprecht, Göttingen, 1966.

Kormendy, C. G., Bender, A. D.: Experimental modification of the chemistry and biology of the aging process. *J. Pharm. Sci.*, 1971, **60**, 167—180.

Kormendy, C. G., Bender, A. D.: Chemical interference with aging. *Gerontologia*, 1971, **17**, 52—64.

Kormendy, C. G.: Aging: can research do something about it?—Pharmaceutical implications. *Proc. 2. Giessener Symposium über experimentalle Gerontologie*, March 16—17th, 1973, Giessen, Germany (in the press)

Lewis, C. M., Tarrant, G. M.: Error theory and aging in human diploid fibroblasts. *Nature*, 1972, **239**, 316—318.

Liu, R., Walford, R. L.: The effect of lowered body temperature on lifespan and immune and non-immune processes. *Gerontologia*, 1972, **18**, 363—388.

Orgel, L. E.: Ageing of clones of mammalian cells. *Nature*, 1973, **253**, 441—445.

Parke, D. V., Rakim, A., Walker, R.: Effects of ethoxyquin on hepatic microsomal enzymes. *Proc. Biochem. Soc.*, 1972, (July 18—20th), 38.

Price, G. B., Makinodan, T.: Aging: alteration of DNA-protein information. *Gerontologia*, 1973, **19**, 58—70.

Pascal, G.: Effets métaboliques d'un additif alimentaire à propriété antioxygène: le BHT. *J. Physiol. (Paris)*, 1971, **63**, 260A—261A.

Walford, R. L.: "The immunologic theory of aging." Munksgaard, Copenhagen, 1969.

Flavour, Taste, and Smell

by A. T. James; Unilever Research Colworth/Welwyn Laboratory, Bedford

Introduction

THE acceptability of a foodstuff to the consumer is dependent on several factors of which the colour, texture, flavour, nutritional value, and price all play a part. These are all parameters that the food manufacturer can use to improve the appeal of the product. With the growth in manufacture and the use of fabricated foods, the availability of natural flavours is so limited that there is insufficient natural material adequately to flavour all food products. There has thus never been such a need to understand the mechanisms of flavour formation in natural foods so that adequate methods for flavour retention, improvement, and enhancement can be established.

'Flavour' has been defined as a combination of those characteristics of any material taken in the mouth perceived principally by the senses of taste and smell but also by the pain and tactile receptors in the mouth as received and interpreted by the brain.

Taste Sensation

The taste sensation is often classified for simplicity into four primary groups: salt, sour, bitter, and sweet to which are sometimes added metallic, alkaline, and astringent. This classification is clearly an oversimplification but these basic tastes are undoubtedly the important ones. These sensations are detected by the taste buds mainly grouped in papillae on the surface of the tongue. Within each taste bud there are 10—15 single taste cells. It has been shown that a single taste cell can respond to more than one taste sensation; however, certain areas of the tongue do respond more readily to specific tastes. For example the sweet sensation is more easily detected at the tip, the bitter at the back, sour at the sides, and saltiness on the tip and at the sides. Thus a child will try to drink an unpleasant bitter medicine without it touching the back of the tongue whereas a tea or wine taster will swirl the beverage over the surface of the tongue to obtain the total impression.

The sensitivity of the tongue varies widely to different taste sensations. This is illustrated by the following taste threshold data.

	Stimulus	Average Threshold Concn.
Salt	sodium chloride	0.25%
Sour	hydrochloric acid	0.007%
Bitter	quinine	0.00005%
Sweet	sucrose	0.5%

These figures represent average threshold data and considerable variation occurs between individuals. It is also found that a low threshold for one sensation is not always accompanied by a low threshold for the others.

Attempts have been made to correlate the primary taste qualities with chemical structure. Although some success has been achieved with sour and salty substances, no correlation has been achieved with bitter and sweet effects. Thus the halide salts of the alkali metals (as well as other inorganic salts) all have a characteristic salty taste of which sodium chloride is perhaps the strongest. Sourness is achieved with compounds possessing an ionisable hydrogen ion. Thus both organic and inorganic acids have a sour taste. However, compounds exhibiting sweet and bitter tastes do not fall into a few chemical classes and indeed the same chemical class may contain substances that are both bitter and sweet.

e.g.

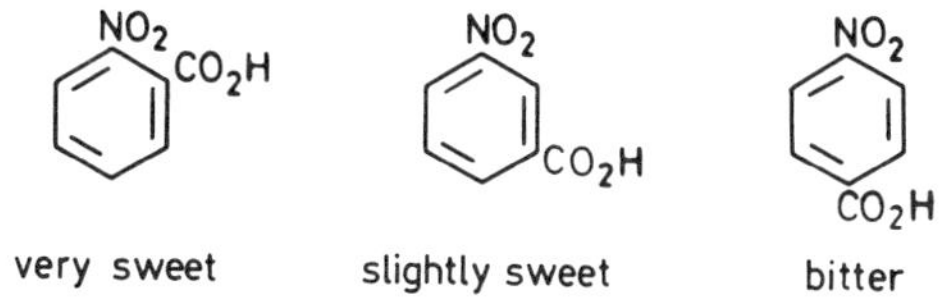

very sweet   slightly sweet   bitter

Sometimes only minor chemical changes are necessary to affect taste quality of a molecule.

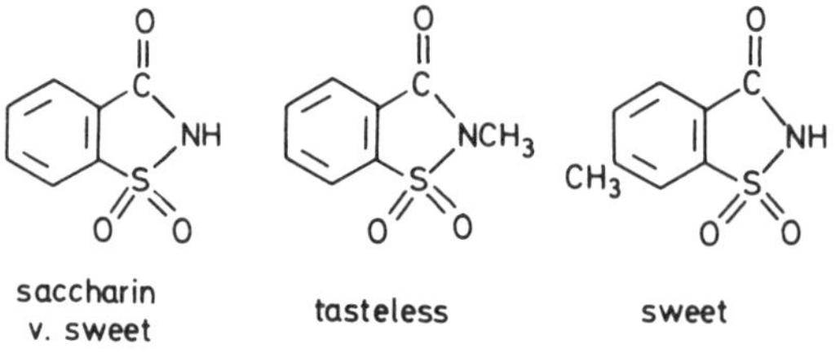

saccharin
v. sweet   tasteless   sweet

Odour Sensation

The smell or aroma of a foodstuff is perhaps the most important single contributor to the characteristic flavour. We all know that an apple and an onion taste very much alike when the nasal cavity is blocked. Odorous compounds must be volatile to be carried through the nasal cavity to the olfactor receptors and therefore there is probably a molecular size limit to compounds which possess a smell. It is generally regarded that molecules containing more than 20 carbon atoms are unlikely to possess an aroma.

Many odorous compounds can be smelled at very low concentrations (*e.g.* the detection threshold for 3-isobutyl-2-methoxypyrazine is claimed to be 1 part in 10^{12} in water). Thus minor impurities in odorous chemicals can have a significant effect on the overall odour character and this has been one of the problems in achieving a satisfactory correlation between chemical structure and perceived odour quality. At present only limited success has been achieved in classifying odour types.

It can easily be demonstrated that compounds with different chemical structure can possess similar smells, *e.g.* different classes of musks:

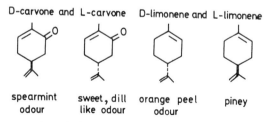

musk xylene   muscone   isochroman type

Equally, compounds with similar structures are claimed to possess different smells. Many pairs of stereoisomers appear to possess different smells, for example:

D-carvone and L-carvone   D-limonene and L-limonene

spearmint   sweet, dill   orange peel   piney
odour   like odour   odour

Although improved techniques such as zone refining have been used to purify selected flavour chemicals more studies need to be carried out on *very* pure materials.

The current theories of olfaction suggest that the size of a molecule is more important than specific functional groups. It is generally assumed that the initial mechanism of olfactory receptor stimulation depends on the adsorption of the odorous molecule at the receptor surface and this will depend both on the general shape of the molecule as well as the type of functional groups.

Davies 'penetrating and puncturing' theory of olfaction[1] has attempted to correlate the cross-sectional area of a molecule and the free energy of adsorption at an oil-water interface with the odour type. The free energy of

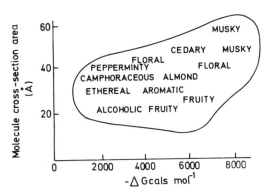

Figure

adsorption will take into account a number of molecular parameters including shape, polarity, functional group type, and location. The odour character of a molecule will then be given by its position on the two-dimensional plot (see Figure). The theory can explain why two molecules with completely different structures can smell alike, providing their cross-sectional area is similar.

Identification of Flavour Components

It has already been suggested that the aromatic components are the most important contributors to characteristic flavour in foodstuffs. The total amount of flavour material in a foodstuff is usually well under 0.1% and this is composed of many constituents representing different chemical classes. Additionally, many of these important constituents will be present at extremely low levels. The alkyl-methoxypyrazines have been estimated to be present in peas at levels between 1×10^{-11} and 1×10^{-12}. Therefore the first stage in flavour analysis is to extract and concentrate the appropriate flavour materials to a level where they can subsequently easily be fractionated and separated. Concentration factors of at least 10^4—10^8 are necessary.

Isolation and Concentration

Methods used for extracting flavour constituents usually involve vacuum distillation procedures or extraction with a volatile solvent which is subsequently removed by distillation. Molecular distillation can be used to advantage for isolating the volatile substances from lipid-based materials. Zone refining and freeze concentration techniques have been used for concentrating materials which are likely to undergo chemical changes when heated and also to concentrate the non-volatile flavour components.

Separation

The flavour concentrates obtained by the previous methods are usually extremely complex mixtures containing perhaps 100—300 components depending on the nature of the starting material. Many different chemical classes will be represented and therefore there is some benefit in fractionating the material into acidic, basic, phenolic, carbonyl, and neutral fractions prior to further high-resolution separation.

Quite often gas chromatography will be used as the final separating technique for volatile flavour fractions; however the technique does have a number of shortcomings when applied to flavour research. The technique can only be applied to those components which are volatile; high-molecular-weight and polar high-boiling components run slowly on GLC columns unless suitable derivatives are formed. Additionally, under adverse conditions, thermal degradation or molecular rearrangement of flavour compounds can occur and many important aromatic compounds may be too unstable to separate. The sensitivity of the normal detection systems is in most cases not comparable with the sensitivity of the human nose—thus, often components can be detected by the nose which produce no detector response.

For these reasons there has been considerable interest recently in the use of high-resolution liquid chromatography which is advancing rapidly as a separating technique mainly due to a better understanding of the factors that affect peak broadening in liquid columns. Liquid chromatography has a number of advantages over gas chromatography for heat-sensitive and higher-molecular-weight (>300) compounds. The column packing can be modified so as to minimise chemical reaction with flavour materials. There is no limit to the molecular size or polarity of the material under investigation, and therefore chemical derivatisation is unnecessary and by using gradient elution one can use an almost infinite number of mobile phases and so maximise the specific resolutions.

Identification
Flavour components are usually identified using both chemical and physical measurements. The spectroscopic methods, including mass spectrometry, infrared, n.m.r., and u.v. spectroscopy have played an important role in elucidating the structures of new flavour compounds.

The mass spectrometer is probably the most convenient and most sensitive physical method available for identifying flavour components, particularly when coupled directly to a gas chromatograph through an interfacing system. With fast scanning spectrometers it is possible to obtain several spectra on one peak and thus identify incompletely resolved components.

Normal infrared spectrometers have very slow scan speeds and cannot efficiently be coupled directly to a chromatograph; therefore it has traditionally been necessary to trap each fraction manually from the chromatographic column prior to obtaining spectra. However, in recent years fast scanning vapour infrared spectrometers have been developed which can scan the spectrum in 0.5 s. Using this type of spectrometer the chromatograph effluent can be scanned continuously as with GLC/MS systems, or alternatively the vapour sample can be isolated and spectra accumulated by computer methods to increase the signal/noise ratio with a consequent enhancement of the spectrum.

Nuclear magnetic resonance spectroscopy is probably the most useful single method for elucidating the structure of flavour compounds, but unfortunately it is the least sensitive of the physical methods. The use of improved Fourier transform systems which allow rapid accumulation of spectra have reduced the sample requirement to under 50 μg.

Usually, complete identification of a flavour component will require the use of several different measurements and final elucidation of the structure will be complete only when the synthesis of the component has been achieved and appropriate comparisons made.

Formation of Flavour Compounds
The components for the flavour of a natural foodstuff are normally derived from the basic carbohydrates, proteins, and fats within the foodstuff by various routes:—

(i) *By biosynthesis in the fruit or vegetable*
One example is the formation of terpenoid materials which occur extensively in fruits, vegetables, herbs, and spices. Most essential oils are composed almost entirely of terpenoid materials. For example, lemon oil is composed

mainly of monoterpene and sesquiterpene hydrocarbons as well as monoterpene aldehydes, alcohols, and esters.

The biosynthesis of the terpenes involves acetyl coenzyme A and follows
a route *via* the mevalonic acid pyrophosphate to geranyl pyrophosphate.
This is the starting point to all the monoterpenes.

Farnesyl pyrophosphate is the starting point to the sesquiterpenoids.

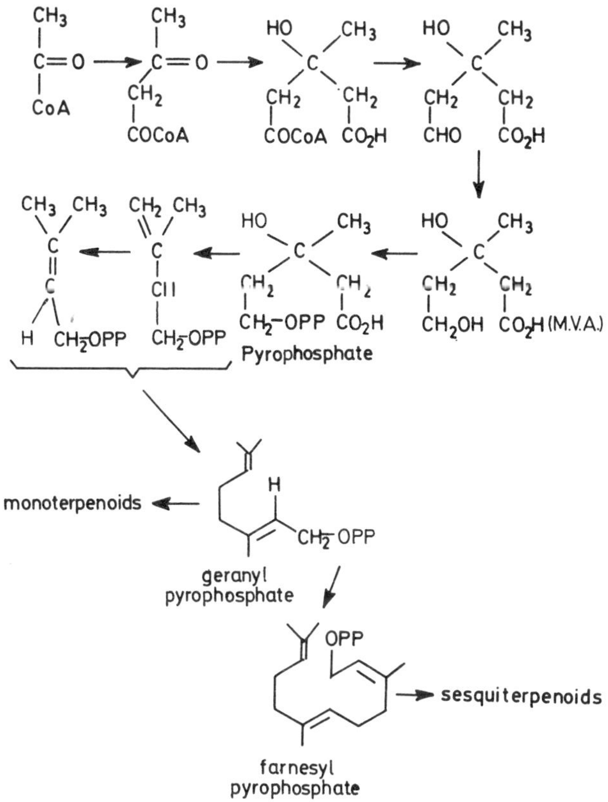

(ii) *Changes after maturity*

When a foodstuff reaches maturity, flavour changes will still occur. For
example if fat is present in the food, lipid oxidation may take place (even
in the absence of enzymes). A low degree of autoxidation can often produce
more acceptable flavour; however if these reactions proceed too far and high
levels of autoxidation products form, the foodstuff can develop 'off-flavours'.
Many of the aldehydes found in dairy products are derived by lipid autoxida-

dation. It is generally accepted that the hydroperoxides of unsaturated fatty acids decompose by a free radical mechanism to the alkoxy-radicals which further degrade to aldehydes. For example linoleic acid forms both a 9- and a 13–hydroperoxide which decompose to give deca-2,4-dienal and hexanal, respectively.

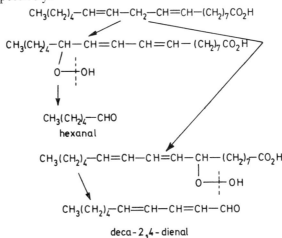

deca-2,4-dienal

(iii) *Flavour development during processing*

A distinction is necessary between processes which derive flavour as a result of enzymic reactions and those where the material is cooked in order to generate flavour.

(a) An example where flavour is derived as a result of enzymic action occurs when a vegetable is chopped or even masticated in the mouth and enzymes are brought into contact with the substrate. For example the *allium* species (onion, leek, garlic) possess mono-, di-, and tri-sulphides as aromatic constituents. These characteristic aroma constituents are absent in the intact tissue and are derived enzymatically when injury occurs by chopping. The allyl derivatives appear to be associated particularly with garlic, whereas the propyl and propenyl derivatives are associated with onion. The substrate *S*-allylcysteine sulphoxide is derived from cysteine (see following page).

(b) Many foods require heating before the characteristic flavour develops. Some typical examples would include the oven baking of bread, roasting coffee, boiling potatoes, and frying of meat. The formation of bread flavour is undoubtedly complex as it involves flavour formation derived both during the fermentation stage as well as during oven baking. However the most

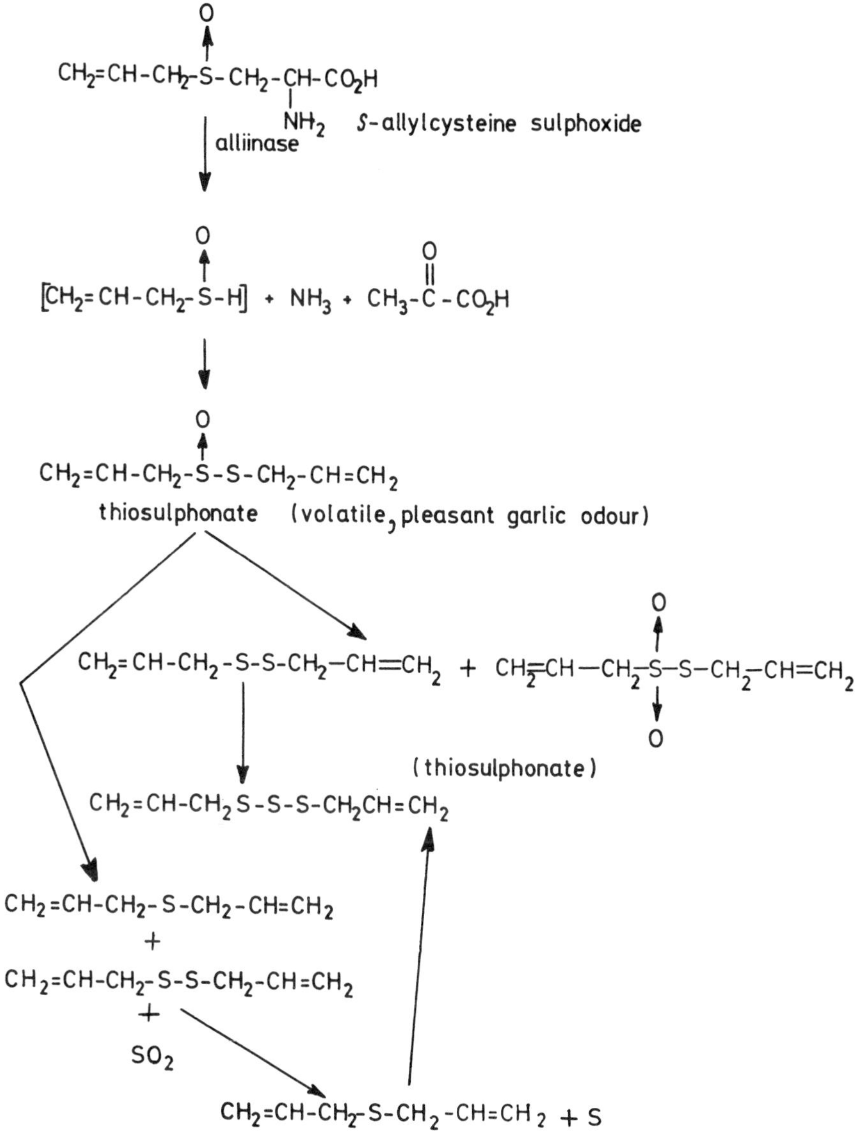

$$CH_2=CH-CH_2-\overset{\overset{\displaystyle O}{\uparrow}}{S}-CH_2-\underset{\underset{\displaystyle NH_2}{|}}{C}H-CO_2H \quad \textit{S}\text{-allylcysteine sulphoxide}$$

alliinase

$$[CH_2=CH-CH_2-\overset{\overset{\displaystyle O}{\uparrow}}{S}-H] \; + \; NH_3 \; + \; CH_3-\overset{\overset{\displaystyle O}{||}}{C}-CO_2H$$

$$CH_2=CH-CH_2-\overset{\overset{\displaystyle O}{\uparrow}}{S}-S-CH_2-CH=CH_2$$

thiosulphonate   (volatile, pleasant garlic odour)

$$CH_2=CH-CH_2-S-S-CH_2-CH=CH_2 \; + \; CH_2=CH-CH_2-\overset{\overset{\displaystyle O}{\uparrow}}{\underset{\underset{\displaystyle O}{\downarrow}}{S}}-S-CH_2-CH=CH_2$$

(thiosulphonate)

$$CH_2=CH-CH_2\,S-S-S-CH_2CH=CH_2$$

$$CH_2=CH-CH_2-S-CH_2-CH=CH_2$$
$$+$$
$$CH_2=CH-CH_2-S-S-CH_2-CH=CH_2$$
$$+$$
$$SO_2$$

$$CH_2=CH-CH_2-S-CH_2-CH=CH_2 \; + \; S$$

Formation of Selected Flavour Materials in Foodstuffs

	Flavour compounds	Precursors	Route
Fruit	Esters, terpenoid alcohols, aldehydes, furans	Acetate	Biosynthesis
	Aldehydes, ketones, esters	Fatty Acids	Biosynthesis
Meat	Sulphur compounds, pyrazines, aldehydes, ketones	Amino-acids/ sugars	Heating
Vegetables	Terpenes, phenols	Acetate	Biosynthesis
	lactones, aldehydes	Fatty acids	Biosynthesis
	Pyrazines, sulphur compounds	Amino-acids/ sugars	Enzymes/ heating
Diary flavours	Fatty acids, aldehydes, ketones, lactones	Fatty acids	Enzymes Heating

significant flavour compounds arise during oven baking and crust formation as many of the volatile fermentation products are lost during baking. It is now well known that the browning flavour reactions are due principally to Maillard reactions.

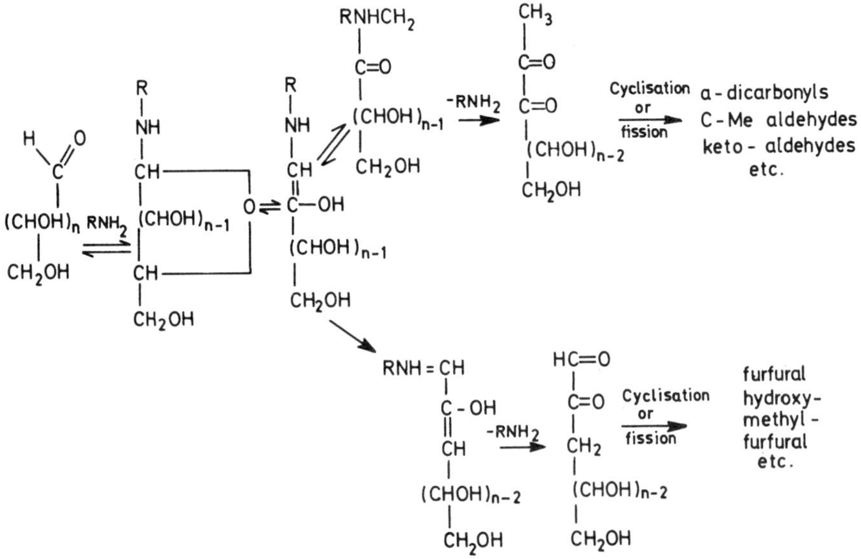

Previously it was considered that the crust formation was a caramelisation process; however caramelisation reactions (involving only sugars) normally require high temperatures whereas Maillard reactions (involving both reducing sugars and amino-acid groups) occur at lower temperatures. The Maillard reaction is represented generally as on page 132.

These products can undergo cyclisation or fission to produce a wide range of carbonyl compounds. The complex melanoidins are produced by polymerisation. Also further reaction of amino-acids by the Strecker degradation produces a series of aldehydes. The amino-acid loses both the amino- and carboxyl groups yielding the corresponding aldehyde with one less carbon atom, *e.g.*

	Amino-Acid		Aldehyde	
Alanine	$CH_3 \cdot CH \cdot COOH$ $\qquad\ \ \|$ $\qquad NH_2$	$\longrightarrow$	$CH_3 \cdot CHO$	Acetaldehyde
Leucine	$(CH_3)_2 \cdot CH \cdot CH_2 \cdot CH \cdot COOH$ $\qquad\qquad\qquad\quad \|$ $\qquad\qquad\qquad NH_2$	$\longrightarrow$	$(CH_3)_2 \cdot CH \cdot CHO$	2–Methyl- butanal
Methionine	$CH_3SCH_2 \cdot CH_2 \cdot CH \cdot COOH$ $\qquad\qquad\qquad\quad \|$ $\qquad\qquad\qquad NH_2$	$\longrightarrow$	$CH_3SCH_2 \cdot CH_2 \cdot CHO$	Methional
Threonine	$CH_3 \cdot CH \cdot CH \cdot COOH$ $\qquad\ \ \| \quad \|$ $\qquad HO \quad NH_2$	$\longrightarrow$	$CH_3 \cdot CH \cdot CHO$ $\qquad\ \ \|$ $\qquad OH$	2–Hydroxy- propanal

(iv) *Flavour Enhancers and Modifiers*

Flavour enhancers or potentiators have been utilised to maximise the flavour of natural foodstuffs—indeed many flavour enhancers occur naturally in foodstuffs. Salt, sugars, and amino-acid derivatives all exhibit some flavour enhancing effect. The most significant amino-acid derivative is the monosodium salt of glutamic acid (M.S.G.) which was first isolated from seaweed about 65 years ago in Japan. (Seaweed has been used traditionally in Japanese cooking for many centuries.) M.S.G. has outstanding flavour enhancing properties particularly in meaty and broth-like flavours and can be used to replace beef extract.

More recently the 5′-nucleotides have been introduced as flavour enhancers of which the disodium 5′-inosinate and the disodium 5′-guanylate are the most commonly used. They are produced by enzymatic degradation

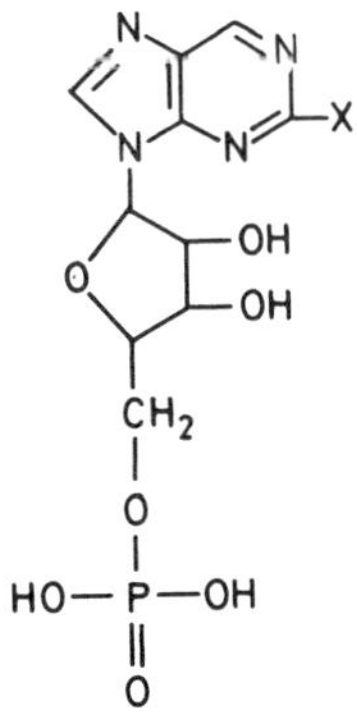

X = H, inosinate
X = NH$_2$, guanylate
X = OH , xanthylate

of ribonucleic acid. Although it is claimed that these ribonucleotides have enhancing properties of their own, there is evidence that they are probably acting synergistically on M.S.G. (present naturally or artificially added to the foodstuff). This is illustrated by the following *taste threshold* data:—

Solvent	Threshold Level (%)		
	Disodium 5′-inosinate	Disodium 5′-guanylate	Monosodium L-glutamate
Water	0.012	0.0035	0.03
0.1% monosodium L-glutamate	0.0001	0.00003	—
0.01% disodium 5′-inosinate	—	—	0.002

The nucleotides appear to be most effective in meat products and also certain vegetables, tomato juice, and instant coffee.

Flavour enhancers which are effective in fruit flavours, wines, and chocalate are maltol and its homologue ethylmaltol. Maltol occurs naturally in coffee and cereals. At the level of usage for flavour enhancement these materials add no flavour of their own.

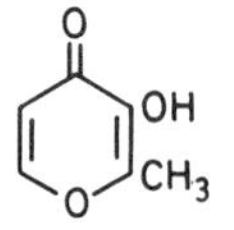

maltol 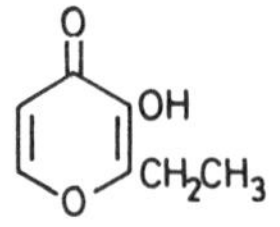   ethyl maltol

Equally important should be materials that are capable of modifying and inhibiting flavour; however these have been less exploited than flavour enhancers. Examples include the well known miraculous fruit which removes the sour taste of a food and replaces the sensation with sweet taste. Beidler claims that it can transform the taste of a sour lemon into that of a sweet orange. Likewise gymnemic acid is capable of removing completely the sensation of sweetness and although the bitter sensation is also reduced, the salt and sour sensation are unaffected.

There is considerable scope for studying and defining other materials that have enhancing as well as inhibitive properties in order to maximise the desirable flavour properties of foods as well as to reduce off-flavours.

The Role of the Flavourist

The discussion so far has dealt only with the function of the chemist and the biochemist but an absolutely critical role is played by the flavourist. Chemical knowledge and biochemical understanding of the components of a natural flavour are helpful to, but not absolutely essential, for the person who compounds the flavour by skilled mixing of a number of components monitored by tasting. Such mixing requires great skill, experience and high flavour/odour sensitivity because it is dependent on the final physiological response of that individual. The skill is thus more of an art than a science and has to be taught to preselected individuals by existing flavourists.

The flavour has to be compounded to give the desired response after inclusion in the foodstuff and after processing and/or cooking, distribution, storage, and sale. Hence a particular flavour cocktail may require extensive modification to suit a different use. The flavourist carries in his/her head a memory of the flavour and/or odour of great ranges of pure chemicals and natural mixtures of many origins. Compensation has to be made for the variability of natural components particularly those imported from countries having no exact quality standards or reproducible extraction processes.

Collaboration between the food processor and product developer and the flavourist is essential for successful application of a flavour to a foodstuff. Most of the younger flavourists and some of the older flavourists are now as familiar with all forms of chromatographic and spectroscopic analysis as they are with flavour compounding.

Safety in Use

The flavour compounder is involved not only in utilising the physiological responses of taste and odour of a compound but also its physiological safety. Although it is commonly thought that 'natural' flavours are safe and synthetic ones potentially dangerous, there is no automatic guarantee of safety in either case.

The general public still does not understand the essential chemical composition of foodstuffs and of those substances, such as flavours, contained in them but having no direct nutritional significance. The fact that no individual food component can be safely consumed above certain levels is still not appreciated so that concepts of safe limits of consumption for food additives is thought to be 'unnatural'.

Since there are now insufficient amounts of 'natural' flavours available the flavour compounders have perforce often to use synthetic compounds. In most cases these are either identical to the natural substances or are chemically similar and hence the emphasis on the identification of the components of natural flavours. Like any other food additive the total possible intake even under conditions of excessive consumption has to be estimated and safe limits defined correspondingly. Unilever uses flavours which have passed through a Clearance System (Philp, 1974)[2] where they are assessed on a basis of their chemistry, natural occurrence, level of use and safety data.

The governmental lists of flavour/odour substances allowed to be added to food products vary from country to country. In any of the 'advanced' countries the time and cost of research to support the submission of a new compound is so great that the emphasis in flavour research has shifted in favour of simulation of 'natural' processes.

The development of new and improved flavours now requires a combined effort by the flavourist, the organic chemist, the analytical chemist, the biochemist, the food technologist, and the toxicologist. Only in this way can we guarantee the public attractive, nutritious, and safe foodstuffs.

Acknowledgments

My thanks are due to Mr M. D. D. Howlett (Food Industries Limited, Bromborough) for the preparation of much of this manuscript.

Bibliography
1 J. T. Davies, *J. Theor. Biol.*, 1965, 8, 1.
2 J. McL. Philp, *Proc. Roy. Soc.*, 1974, *B*, 185, 199—208.

Chemistry, Colour, and Communication

by G. F. Duffin; Minnesota 3M Research Ltd., Pinnacles, Harlow, Essex.

ONE of the most surprising things about technological progress is the degree to which we have become accustomed to it. The picture shown in Plate 1 would have been mere fancy less than two decades ago but it is one particular aspect of it to which I wish to draw your attention. How long is it, I wonder, since you picked up a magazine which was printed twenty years ago, or come across, while tidying a neglected corner of your home, a popular book printed before the War? Indeed, it would surprise most of us to look at the postcards which we sent from our holidays only twenty years ago to be reminded of the muddy, toneless, sepia prints which were intended to convey to our friends what a beautiful place we were visiting. I refer of course to the fact of colour.

Turning back to the space exploration, I imagine that most of us must have watched a splashdown at the end of one of the American moonshots, and this we can do in our own homes and see a full-colour realistic reproduction from the other side of the world conveying the whole atmosphere and impact of the event. Yet I remember that when I showed colour slides based upon Ilford Ltd's immediate post-war colour film technology, I had to explain to the audiences that these were truly colour photographs and not produced by hand colouring of black-and-white pictures.

It is a truism to say that we have become used to colour in every medium of communication—even the newspapers include colour pictures, the glossy magazines become glossier and glossier and convey more realism, and it is fair to say that we are no longer prepared in any field of human communication to rely upon black-and-white images. I shall discuss a number of areas in which chemistry has played a large part in this particular enrichment of the quality of life today and show that it is a part which we have all come to accept and, I submit, would feel the much poorer were we without it.

I suppose it is natural to turn first of all to colour photography and by colour photography one's initial thought must be for films as used in the snapshot camera. The physical basis of modern colour processes was laid down 100 years ago but it is only since the Second World War that colour photography, with a realism which today appears totally acceptable, has been available to the ordinary person. I think there are three aspects of this on which such progress has been made in the last decade or so. There was a time when colour films were very slow, the ones which I referred to earlier were 6 ASA or thereabouts, whereas now, the progress in the preparation of the sensitive silver-halide crystals and the tremendous strides in the balancing of the chemical processes involved in the development of the colour image, have given us films which can be processed up to 1000 ASA. (See Plates 2—4.) Most colour materials today still rely upon the chromogenic

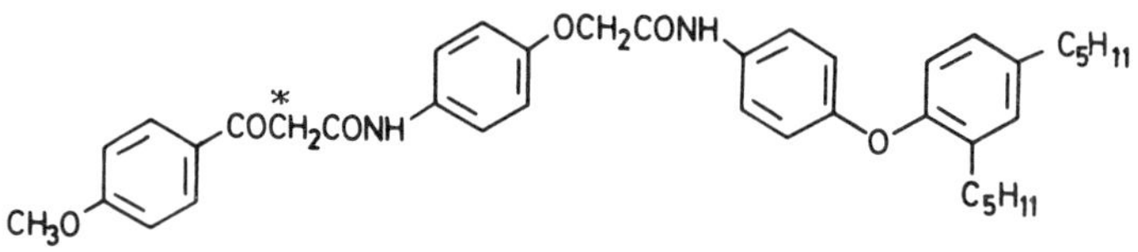

(I)

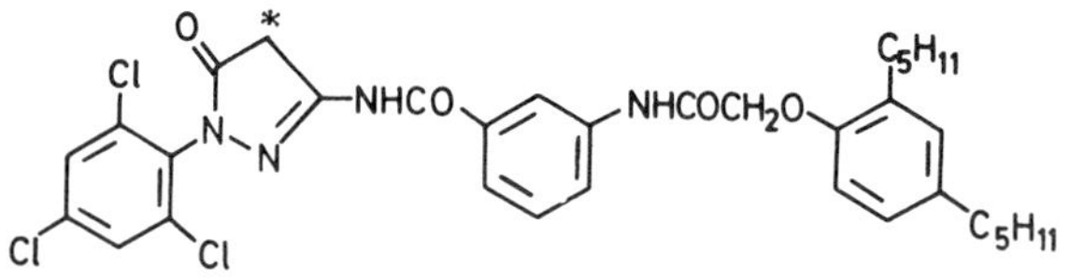

(II)

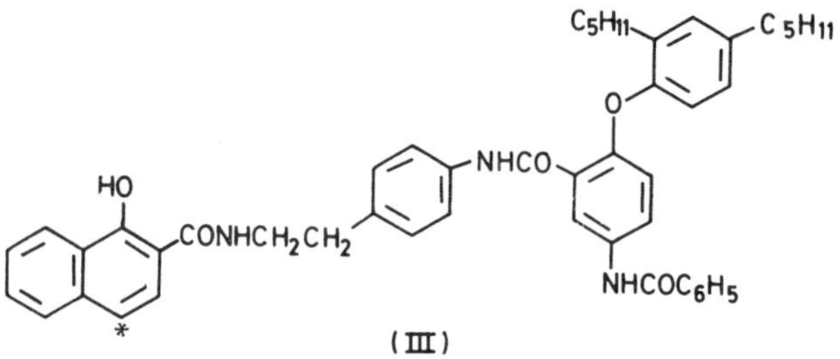

(III)

development system and here a diethyl-p-phenylenediamine developer reacts with one of three classes of compounds to give, from the same oxidised developer molecule formed in the development process, dyes absorbing the three primary colours. Thus an acetoacetanilide (I) gives a yellow image, a pyrazolone (II) gives a magenta image and a naphthol (III) derivative gives a cyan image. The complicated molecules used in this process, as will be obvious from the slides, are tailored in a manner required by the make-up

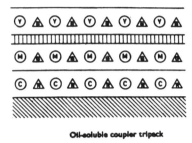

Oil-soluble coupler tripack

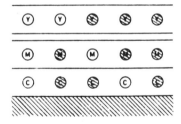

Oil-soluble coupler — final state

Figure 1.

of the colour film. The formation of a true colour image requires that the yellow dye be formed in a blue-sensitive layer, the magenta in a green-sensitive layer and the cyan in the red-sensitive layer. This being primarily an exposition of chemistry, I do not propose to discuss the detail of the physics at this point. However, the sole formation of the correct dye in the appropriate layer is ensured by solution of these complicated colour-forming materials in a solvent such as dibutyl phthalate, the whole then being emulsified in the dispersed phase in the silver-halide aqueous layer. This is shown in

stylized form in Figure 1. The bulky groups in these colour-coupler molecules, as they are called, are designed to give the correct physical properties
to the final layer. Practically all conventional colour photographic films
today use this technology. All such materials form their coloured images
by a similar reaction with oxidized developer as shown in a simplified
manner:

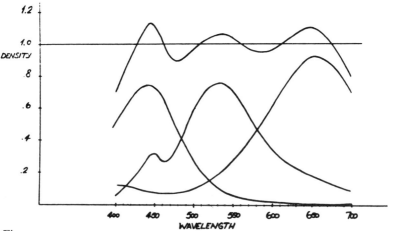

   In spite of the attention that the chemist applies to choosing the colour-
forming substances so that the final dyes are as ideal as possible, the magenta
and cyan materials always possess spurious absorptions at shorter wave-
lengths. These are shown in Figure 2. While any adverse effects of these
secondary absorptions can easily be overcome in reversal films, such as
Kodachrome designed for slide use, they do introduce serious difficulties

Figure 2.

Courtesy NASA

PLATE 1   Earth rise over lunar horizon

PLATE 2

PLATE 3

PLATE 4   Plates 2–4: Colour slides taken on Ektachrome (high speed) in available light

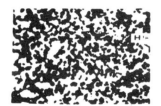

Silver developed in Dmax area of emulsion containing conventional coupler.

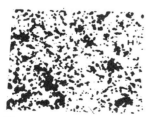

PLATE 5

Silver developed in Dmax area of emulsion containing DIR coupler.

PLATE 6   Scotchlite Reflective Sheeting signs at night

when employed *via* a colour negative for prints. The high quality of the colour prints available to the amateur photographer today stems from another invention of the chemist. Instead of a straightforward, *e.g.* pyrazolone, coupler, one is used containing an inherently coloured group in the coupling position:

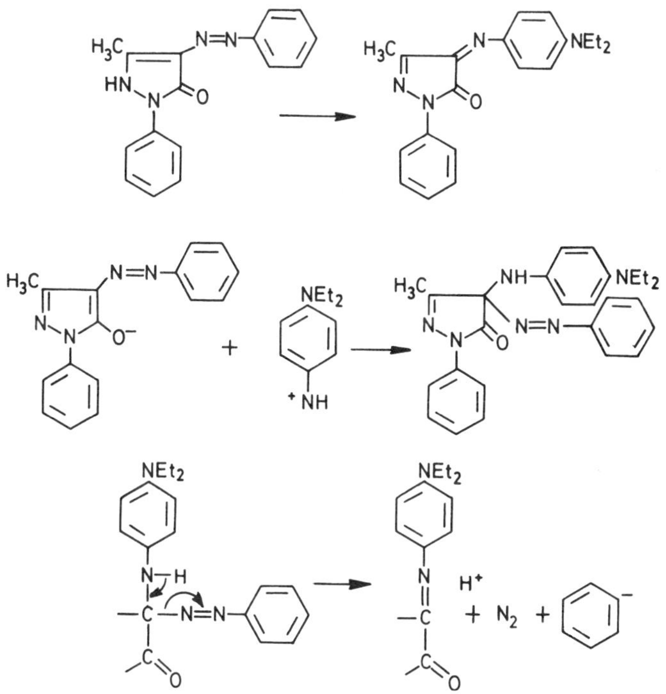

When reaction occurs with the oxidized colour developer, the usual azo-methine dye image is formed, whereas in the non-image areas the yellow colour due to the azo-group remains. This means that the image-wise effect of the spurious blue absorption of the pyrazolone dye is balanced by a counter-imagewise absorption of the unreacted coupler and thus there is no image to degrade the negative. The resulting print from this material on to a positive paper for the final print is therefore much purer.

Another interesting use of a similar concept is a method called developer inhibitor release coupling. In this case, a similar doupler (IV) contains a group in the coupling position (9) which on reaction gives a perfectly normal

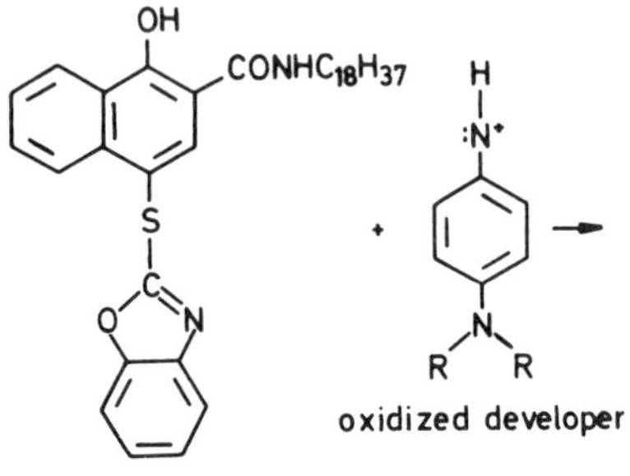

oxidized developer

(IV)

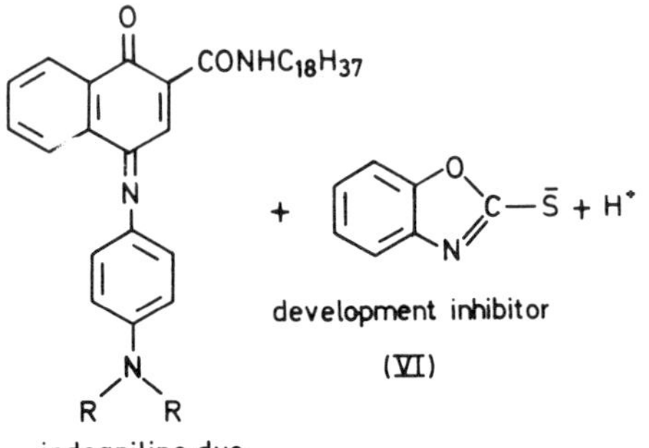

development inhibitor

(VI)

indoaniline dye

(V)

dye (V), expels a molecule, such as the mercapto-compound (VI) shown, which is itself a powerful inhibitor of the development of silver-halide crystals. This therefore slows further development and prevents the spreading of the image, in a given development area, which would give a 'grainy' effect to the final image. This effect is shown dramatically in Plate 5. This technology thus permits less grainy pictures to be obtained and is the basis for the present small-format amateur cameras which are relatively cheap, easy to use, and convenient to carry around.

There is a third area of technology which has made dramatic progress in the last two years or so; I refer to the Polaroid Instant Colour Print Process. The make-up of this is shown in stylized form in Figure 3. The camera is designed optically to achieve a transposition of left to right and the film is exposed as shown in the slide. On expressing the 'picture' from the camera the pod is broken to spread its contents very evenly through the gap between the front receptor layer, at that point transparent, through which the image was formed, and the complex light-sensitive element beneath it. Development then takes place and, as shown in Figure 4, the appropriately

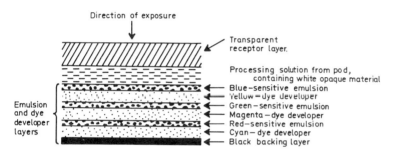

Figure 3.

coloured dye developers diffuse over from the areas in which development *does not* occur but are retained where development occurs. This gives an image in the receptor layer which grows before the eyes of the photographer. In order to protect the originally light-sensitive element from further action the opaquing layer is formed, again by the contents of the pod, in the intervening space and the final picture can be regarded as permanent.

It could well be argued that the contribution of the chemist to colour TV, as dramatic as it is in its impact, is less than in colour photography to which I have just made reference. However, while the process may rely upon the basic physics of colour reproduction, the nature of the phosphor, designed to emit light of the correct colour, must be very carefully chosen. The chemist has been involved in the preparation and purification of rare-earth emitters of very high purity and carefully-chosen crystal size and structure, the solid-state chemist, as well as the physicist, has contributed to this to provide the correct screens for colour TV. The red-emitting material proved to be a problem, but the use of yttrium vanadate doped with europium has provided a red phosphor superior to earlier versions. The green phosphor is based on zinc–cadmium sulphide, and the blue on zinc sulphide.

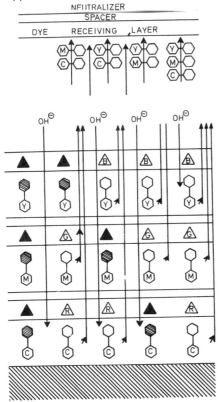

Figure 4.   Polacolor — development

It is not so many years ago since document copying broke on to the market and, in a surprisingly short time, the relatively restricted output of blueprints and the diazo-copying processes gave way to the modern reprographic industry. In turn, the black-and-white document copying revolution is now being followed by copying in colour. I would like to refer to a colour copying process which has been disclosed fairly recently. It is due to the 3M Company and is called 'Color-in-Color'. This is a process which relies upon three images, as usual one of each of the primary colours, which are laid down in turn, but uses a very ingenious system in which full registration for the three successive images is retained throughout the process.

The machine for doing this is shown in Figure 5. The original document is placed in position (1) and then *via* the optical system (2), an image is

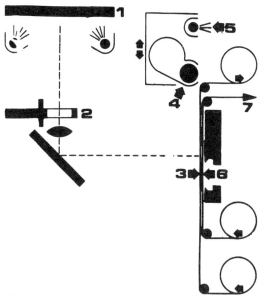

Figure 5.

formed on the front surface of the roll (3). This is electro-image which is then developed by moving down the developing station (4) which deposits a black-absorbing image where the light fell from the original exposure. After the developing station has finished its task, the infrared lamp (5) passes in front of the developed image, the infrared radiation being absorbed and causing the volatilization of a dye coating on the back of the intermediate (3) coated with a yellow dye and depositing the yellow image at those points on the final picture (6). A blue filter had been used for this first exposure of the original and therefore the desired image is formed in the yellow areas of the original picture. The intermediate (3) is then moved one frame-width, the appropriate area being then coated with a magenta dye. The exposure, development, and infrared cycle is then repeated using a green-filter exposure, transferring the magenta parts of the picture to the final gap (6). A third treatment with a red filter for the original and a cyan dye completes the picture.

It will be seen that there are a number of keys to the success of this method, the first one being that the original (1) and the receptor (6) remain optically locked throughout the imaging steps, and thus perfect registration is achieved and also that sufficient volatile dyes of the desired hue are required on the back of the intermediate (3). Very little dye is used and the high tinctorial power of the sytem makes it very attractive in terms of cost. Each of the exposures takes 1—2 s, and the whole process is completed in 30 s.

You will also appreciate that, due to the nature of the method it is possible to produce colour images from black-and-white ones, and also to change the colour of any material by selecting the colour one intends to print. Therefore, from one original, a number of different colour selections can be made with very little trouble.

I have already made reference to the tremendous improvement in quality of the coloured pictures in our magazines, calendars, etc. Apart from the fact that the present advanced state of coloured photography is used as the basis for making these pictures, there have been many chemical advances in printing technology which have given today's wonderful pictures. Colour printing depends upon using a number of different separate images, made from the original full colour picture by separating into the component parts. The picture must then be re-assembled with all the colours, and black as it turns out, in register, and this makes stringent demands upon the individual separation pictures. The only way of achieving satisfactory dimensional stability for the photographic materials used in this process was, for many years, to employ glass plates. These are very inconvenient. However, the introduction of polyester base, that is poly(ethylene terephthalate) or 'Terylene', has given sufficient dimensional stability to films to enable full colour work of the highest quality to be done using otherwise conventional film materials. The other big area of advance has been in the improvement of the printing inks. Technology in this area is a very closely guarded secret but the chemists concerned have produced very dramatic improvements in the gloss effect of these inks and the last five years has seen tremendous strides forward.

There are two other areas, somewhat inter-connected, to which I would make reference in this consideration of colour. We all remember the time when roadsigns were merely painted, or perhaps you can remember the old Franco signs which, if you were lucky, gave a weak glimmer from your headlights and just about identified something dimly at the side of the road. Nowadays however we have the brilliant modern signs and I think it would be as surprising to look back to the old dull white signs as it would to return to black-and-white photography. Today our motorways are lit with signs of various colours and Plate 6 shows a selection of today's coloured signs based upon modern reflective technology and dyestuff applications.

In the daylight too, signs have become brighter and more eye-catching.

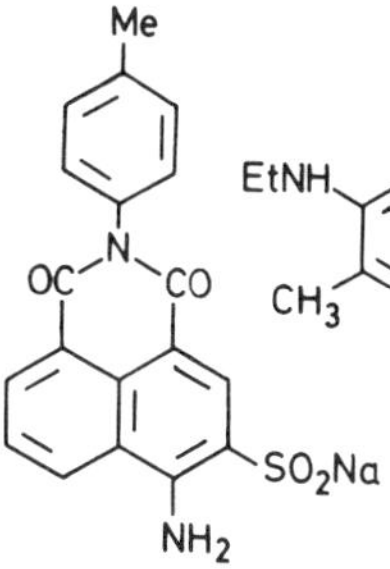

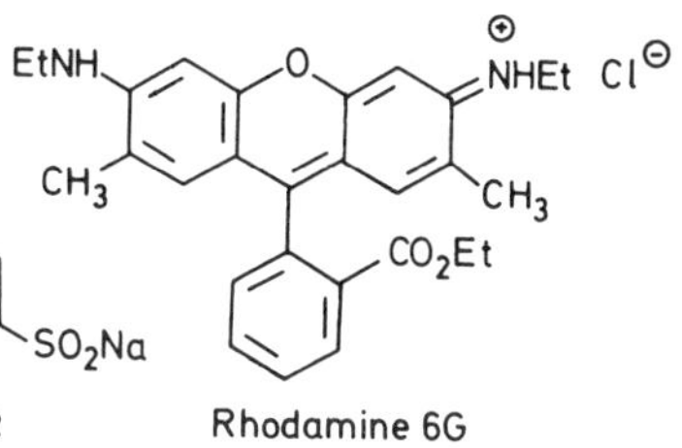

Brilliant Sulphoflavine FF   Rhodamine 6G
(VII)   (VIII)

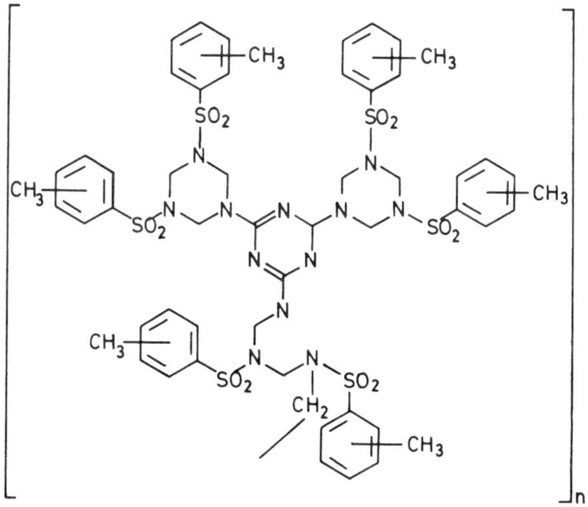

This advance is based again upon new technology derived by the organic chemist. Two of the dyes which are used in this application are compounds (VII) and (VIII) and, although based upon older types of dyestuff, specific examples have been synthetized in order to give the brilliance that we have become accustomed to seeing. However, the application of these dyes is really the key to the tremendous fluorescence which the final coatings exhibit. The dyes are dissolved in a specially designed plastic because the fluorescence of organic molecules is efficient only in solution, usually fairly dilute solution, the object being to stop the dyes being returned to the ground state by any other mechanism than that of a fluorescence emission. The type of plastic used in this is illustrated as (IX). After dissolution of the dye the plastic is then milled and coated as a particle; they are referred to as pigments but this is rather inaccurate. It would be better to call them dispersions of dyed plastic.

We have thus seen many examples in which the science of chemistry has been applied to the use of colour in communications. This has been particularly true of the last twenty years or so in which tremendous advances have been made. It would be difficult indeed to forecast the future but I am sure that colour will continue to play an increasing part in our everyday life and that the ingenuity of the chemist will once again provide the main impetus for advances.

In conclusion, I would like to thank my colleagues within the 3M Company for their help in the preparation of this paper, Kodak Limited for the provision of the Ektachrome slides, Sir James Taylor, F.R.S., for the kind loan of two slides, and Dr. G. de W. Anderson of the Paint Research Station for his assistance with regard to the fluorescent pigments.

Proficiency with Polymers

by C. E. Hollis

Introduction

THE theme to which this paper is supposed to conform is 'Enjoying Life'. There are many contributions which chemistry can make to this desirable experience, but some would say that the theme is a very materialistic one: however, my particular subject, polymers, fits in aptly, since synthetic plastics, fibres, and rubbers are now so widely used as materials from which to fabricate things we need—and some we don't—that they tend to be taken for granted, at least in the so-called advanced communities. If the dreaded, but so far hypothetical, micro-organism (*B. domesdagwacche?*) suddenly began to metabolise and destroy the synthetic polymers in our houses, vehicles, and factories, life as we know it would be completely disrupted, We need only think of a single effect, for example the loss of synthetic insulating materials, in order to conjure up a gloomy or maybe lurid picture. Fortunately, in spite of certain efforts in this direction by those whose preoccupation is with plastic litter, it seems that such a catastrophe is unlikely from this particular cause.

However, the present problems of energy and raw materials supply have shown us that we ought not to become complacent about the continued ready availability of the many useful and sometimes spectacular products based on synthetic polymers. Furthermore, whatever the supply problems of the moment, the polymer industry in the U.K. has to face increasingly direct economic competition in the E.E.C. with some of the most competently professional manufacturers and vendors of polymers in the world. It is vital that we should use our materials to the best advantage: we must be proficient. The O.E.D. defines proficiency as 'advancement towards the attainment of a high degree of knowledge or skill; adeptness; expertness'. How adept are we at making and using polymers? Is our expertise satisfactory? Such questions are likely to become even more important to us than they are at present; if we are not proficient our enjoyment of life will soon be at a lower level than it is now.

Polymers as Materials

It would require a tedious catalogue to cover all the uses of synthetic polymers to which we have become accustomed; it is difficult to find any area, domestic or industrial, where rubbers or fibres or plastics are not used (see Plates 7—9). In many cases polymers are vital components of things which could not be produced at all if we were confined to the use of traditional materials. On the other hand, there are some uses, particularly of plastics, which do not seem to be very desirable from the point of view of enjoying life. In discussing these matters, it will be useful to recall one or two historical facts (Table 1). The use of synthetic polymers originated in empirical

Table 1:

---

1. Origin of synthetic polymers, to replace or extend natural materials, was empirical.
2. New technology developed very fast, retains much empiricism.
3. Success has been due to:
        (a)   cheap raw materials,
        (b)   easy fabrication methods,
        (c)   useful, new combinations of properties,
        (d)   mass markets for products.
4. Polymer properties are still unfamiliar, 'peculiar'.

---

attempts to replace natural materials (*e.g.* natural rubber, bitumen, shellac, drying oils, and natural fibres). The new technology which has arisen in the last one hundred years, although now essentially science-based, necessarily retains a good deal of its early empiricism, more particularly when it comes to fabrication methods.

The economic success of polymer technology depends upon many factors. Cheap chemical raw materials and intermediates have been available; useful combinations of physical properties, not always present in other materials, can be obtained; relatively easy methods of manipulation and fabrication into end products have been found, suitable for mass production techniques; finally, the development of mass markets for these products has been achieved.

Most people have fairly clear ideas about the mechanical properties of traditional materials such as stone, wood, metals, and glass, based on long experience of large and small articles made from them. Synthetic plastics and similar materials are as yet not so well defined in the public mind, partly because they are relative newcomers among usable materials, partly also because they can have so many different attributes. They can be glassy, fibrous, leathery, papery, rubbery; hard or soft, plain or coloured, brittle or stronger than steel. This is all very confusing; after all, metals are almost always metallic, and everyone can recognise a brick. It is also unfortunate, because it would be ultimately to the good of the industry if the user had a much better idea of what he or she should demand of polymeric products. On a more technical level the physical properties of polymers can be generally characterised as 'viscoelastic' (Figure 1). The mechanical behaviour of

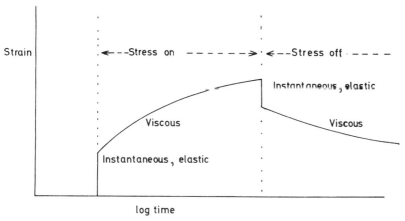

Figure 1. Simple visco-elastic behaviour. (Constant temperature and stress.)

polymers, involving marked temperature-, time-, and load-dependence, can well puzzle a design engineer brought up on the use of metals, normally at temperatures far removed from their melting points.

In the case of traditional materials, individual skills in their use have usually developed into crafts, and crafts into technologies of a sort, long before a scientific approach was possible. Much the same has happened with polymers but the development time has been extraordinarily compressed. Wood, stone, metals, and glass were put to highly ingenious and artistic use several thousand years ago: synthetic polymers have been with us for less than one hundred years. Until the mid-1930s, what may be called polymer prototechnology was much more indebted to craftsmanship, trial and error, and ingenuity than to anything we should call polymer science. Baekeland started to commercialise phenol–formaldehyde resins before there was any agreement about, or indeed much interest in, the nature of macromolecules. Nitrocellulose billiard balls, combs, and collars were well known even earlier, and so were rubber tyres. Viscose fibres were used before there was much detailed knowledge of the structure of cellulose.

Science, both pure and applied, began to affect this prototechnology when the ideas of H. Staudinger, H. Mark, and others began to circulate, and when industrial chemists, mainly in the U.S.A. and Germany, began to realise the commercial possibilities of high-tonnage chemicals from coal and, later, oil. The production of polychloroprene, polystyrene, and poly(vinyl chloride) began in the 1930s when it was seen that such materials could not only replace natural products such as rubber, bitumen, or shellac, but could also be used to make things with improved performance; for example, better electrical insulators, or varnishes with better water-resistance. Many chemists who are still with us will remember this phase, and the associated difficulties of building up a worthwhile market; it is difficult now to imagine how this was done in the absence of cost–effectiveness analyses, market forecasts, and research efficiency studies! The Second World War accelerated the change to a technology proper, as a result of the urgent national needs for large amounts of synthetic rubbers and plastics. Intensive development work was carried out and large-scale polymer production methods were improvised; close contact between research and production people was enforced by necessity, and as a result, the knowledge of polymers had advanced remarkably by 1946. Academic research into polymer problems became fashionable at the same time. Since then, science and industrial practice have both made continuous advances; a great many feedback loops have been established between polymer technology and chemical research, to the advantage of both.

These rather obvious historical points are stressed because there is so much emphasis on 'polymer science and technology' today that it may not be appreciated that the practical and empirical origin of polymer technology still exerts a marked effect on its development; it would be unreasonable to expect that every corner of such a technology should yet be bathed in the light of science.

The Chemist as a Tailor of Macromolecules

The basis for the 'peculiar' properties of polymers is the macromolecule, in which large numbers of small molecular units (derived usually from quite simple organic compounds) are strung together, like beads in a necklace, into long flexible chains which when unconstrained are kinked and coiled into a vast number of configurations. The chains may become linked together to form 2- and 3-dimensional networks. Typical polymeric behaviour is in simple terms due to the gross entanglement which takes place when the chains become long enough; once they contain a few thousand units it is impossible for individual chains to sort themselves out into a crystalline array, hence the natural solid state of the material is that of an amorphous glass. Because of the general disorder in such a material, it will contain internal voids (free volume) and there is room for short lengths or segments of the chains to rotate about in-chain bonds and to move about, if sufficient thermal energy is available. While the relative diffusive movement of whole molecules is thus possible, it takes place very slowly in comparison with movement in materials made up of small molecules. If, at ordinary temperatures, there is enough thermal energy to activate the chain wriggling movements, the material will behave in a rubbery manner; if not, the polymer will tend to be a glassy solid. The internal mobility of polymer molecules under use conditions is then responsible for many of the characteristic properties of polymers; for example, for their tendency to creep under prolonged mechanical stress (which is a nuisance in a flat-spotted tyre or a fractured water pipe), for the highly elastic behaviour of rubbers (which is advantageous when landing an aircraft), and for the local ordering or crystal-

Table 2: Structural Factors in Polymer Molecules

Factor	Behaviour (at about 20°c)
1. FLEXIBILITY—ease of kinetic motion of chain segments (especially rotation)	*Flexible*—rapid conformational changes,—rubbery behaviour *Stiff*—very slow conformational changes,—glassy behaviour
2. REGULARITY—ease of repeated segmental packing	*Regular, symmetrical*—partial crystallisation possible,—disordered crystal or fibrous behaviour *Irregular,.bulky*—remains amorphous,—glassy behaviour
3. INTERCHAIN FORCES—amount of cohesion between chains	*Low* (van der Waals and dipole) *Moderate* (H-bonding) *High* (covalent or ionic) Modify basic properties dictated by (1) and (2). Usually mechanical strength increases with degree of cohesion

lisation in polymers of the right architecture (notably useful to the textile technologist).

Chain architecture is very important, and polymer characteristics can be qualitatively rationalised by considering the balance of three general attributes of the macromolecules indicated in Table 2. These architectural

Table 3: Some Rubbers

Chain unit[a]	Properties[b]
1. $\diagdown C \diagup C{=}C \diagdown C \diagup$	Poly-*cis*-butadiene. $T_g - 108$, $T_m - 11$. Strong, highly elastic. Similar to natural rubber
2. $\diagdown C \diagup C{=}C \diagdown C \diagdown$	*Trans*-Isomer of (1). $T_g - 18$, $T_m - 148$. Resinous like Gutta-percha[c]
3. $\diagdown C \diagup \overset{\overset{Cl}{\mid}}{C}{=}C \diagdown C \diagdown$	Polychloroprene. $T_g - 45$, $T_m - 80$. *trans*-configuration and Cl-substituted, yet is a strong rubber; good age- and oil-resistance[d]
4. $\diagdown C \diagup \overset{\overset{C \quad C}{\diagdown \diagup}}{C} \diagdown$	Butyl rubber.[e] Good age-resistance, low permeability to gases[f]. $T_g - 73$
5. $\diagdown O{-}\overset{\overset{C \quad C}{\diagdown \diagup}}{Si} \diagdown$	Silicone rubber. $T_g - 123$, $T_m - 39$. Good heat- and solvent-resistance
6. $\diagdown S{-}\overset{\overset{F \quad F}{\diagdown \diagup}}{C} \diagdown$	Polythiocarbonyl fluoride. $T_g - 118$, $T_m\ 35$. Tough, resilient, v. resistant to chemical attack
7. $\diagdown C \diagup \overset{\overset{C \quad C}{\diagdown \diagup}}{} C \diagdown$	Polyethylene. $T_g - 125$, $T_m\ 138$. Not a rubber normally, crystallises too readily

[a] H-atoms omitted
[b] Useful rubbers must be elastic at use temperatures (-50 to $50°C$), hence glass temperature, T_g, should be low. Crystalline melting point, T_m, is usually low, but effects of this are complex.
[c] (2) is more symmetrical than (1); crystallisation supresses high elasticity. Compare (7) also.
[d] Cl-substitution in (2) reverses the properties and (3) is an excellent rubber.
[e] Contains small % unsaturated groups to permit vulcanisation.
[f] (4), (5), and (6) all have flexible, low-cohesion chains and are all rubbers, in spite of chemical differences.

factors are oversimplified and are only a rough guide to the physical properties a polymer can be expected to have; thus, polymers which are rubbery
(highly elastic) at ordinary temperatures usually have very flexible chains
showing low interchain forces, while useful fibres would be expected to be
based on highly interactive macromolecules, preferably able to crystallise,
in order to ensure good strength and a melting point sufficiently high to
resist washing and ironing of fabrics made from them. The exact details of
chain make-up are also important in deciding the texture of the solid polymer,
on which so much depends. Tables 3 and 4 give a few practical illustrations
of the effect of the factors shown in Table 2.

Table 4: Some Fibrous Polymers

Chain unit	Fibre		
–CCCCCC–	Linear Polyethylene	T_m 138	
–C$_{10}$CONH–	Nylon 11	T_m 187	Increase in H-
–C$_5$CONH–	Nylon 6	T_m 215	bonding, rise in
–NHC$_6$NHCOC$_4$CO–	Nylon 66	T_m 263	melting point
–NHC$_6$NHCO–p-C$_6$H$_4$–CO–	Polyhexamethylene terephthalamide	T_m 370	Rings in the
–NH–m-C$_6$H$_4$–NHCO–m-C$_6$H$_4$–CO	Poly-m-phenylene isophthalamide[a]	T_m 371	chain increase melting point
–NH–p-C$_6$H$_4$–NHCO–p-C$_6$H$_4$–CO–	Poly-p-phenylene terephthalamide[b]		
–NH–p-C$_6$H$_4$–CO–	Poly-p-benzamide[b]	Sinters >350	

[a] Du Pont "Nomex" fibre; for fire-resistant clothing, etc.
[b] High tensile fibres with high heat-resistance.

Even before these general principles were sorted out, Carleton Ellis, that
encyclopaedist of early polymer technology, coined the phrase 'tailoring the
long-chain molecules',[1] implying that by the careful choice of their chemical
structure, polymers could have specific properties built into them, e.g. high
strength, heat resistance, solvent resistance, or easy processability. Living
cells are the real experts at tailoring, and in the origin of their astonishing
skill lies the mystery of the creation of life; it is no wonder that many
polymer chemists have gone over to molecular biology. Ham-fisted polymer
technologists, confined to the use of a limited number of cheap building
blocks and to rather brutal chemistry, are not so competent as cell enzymes
and have to resort to a good many fitting sessions with their customers.
However, even without much benefit of fundamental research, they de-

veloped, as already mentioned, many plastics, fibres and rubbers (*e.g.* polystyrene, nylon 66, polychloroprene) whose production is now reckoned in hundreds of thousands of tons per annum. With the powerful synthetic and analytical methods available today, supplemented by our very greatly increased theoretical background, we should be able to do much better still; in fact it can fairly be said that chemists are now quite good tailors of industrial polymers.

Texture in Polymers

If we pursue the tailoring analogy a little further it will be realised that the weave and construction of the fabric being cut is likely to be just as important—perhaps more important—than the thread used. The detailed make-up and steric structure of many macromolecules can readily be worked out today by using the various highly developed spectroscopic and other techniques at our disposal. It is now commonplace to determine routinely characteristics, even of commercial polymers, which would not have been accessible ten years ago. Thus, molecular weight averages and distributions in commercial polyolefins are regularly determined by gel permeation chromatography because the moulding performance of these plastics can be correlated with these measurements. Theoretically, such information can be related to the properties of individual molecules, or to assemblies of these in which the individual chains are well separated, as in dilute solutions. In practice it becomes very difficult to relate the properties of polymeric materials in the solid state, or even as viscous melts, to their basic chain structures in a quantitative manner. Certainly in the case of melts, viscosities can usually be related to the weight-average molecular weight of the polymer concerned, by empirical equations, but when it comes to most of the physical properties of solid polymers it is often more useful to think in terms of continuum, rather than molecular, properties; we then enter the realms of physics and engineering rather than chemistry. Nevertheless, a great deal of knowledge has now been gained about supramolecular textures in polymers which throws much light on their mechanical behaviour.

The polyethylenes may be taken as an example. It has already been noted that although linear polyethylene has chains of good flexibility with low intermolecular forces, it is yet not a rubber at ordinary temperatures; this is because the chains have a simple symmetrical structure and segments of them can pack nicely into a crystalline lattice; when this happens to a sufficient extent, the material becomes too rigid to show very much high elasticity. (If the regularity of the chain is broken up by copolymerisation, as in random ethylene–propylene copolymers containing 20–30% mole of propylene, rubbers are in fact obtained; this is the chemical basis of the E.P.D.M. rubbers.) Hence, when molten polyethylene, which is already very viscous, cools and solidifies it forms a mass of interconnected micellar crystallites, which aggregate into rosette-like multi-crystalline sheaves, or spherulites. Quite a lot of amorphous material remains within and between the spherulites so that the solid product, by the time it has cooled down to, say, 20°C has a very complex felted kind of structure. Close examination by low angle X-ray spectroscopy, electron and scanning electron microscopy, and other techniques has shown up further details, in particular that the crystalline regions are made up of chains which have folded back and forth on themselves many times, giving a concertina-like crystal structure. Many chains are involved in several crystalline regions, which are thus linked together chemically. The crystalline regions confer a much greater rigidity on polyethylene than it would otherwise have at ordinary temperatures. The size of the concertinas, of the spherulites they make up, the degree of their orientation (the direction in which they point), and of the amount of amorphous material which has not crystallised depends on many factors; notably on the rate of cooling of the melt, the presence or absence of nucleating impurities and on whether or not the solidifying material has been static or forced to undergo flow (e.g. in an injection moulding machine) while it was cooling. Flow under extreme shear conditions can in some cases cause different type of crystallisation in which the crystal habit is different and the chain molecules are aligned parallel to each other in bundles.

What is the practical result of all these complications? The answer is that the degree and type of crystallinity has a considerable effect on many mechanical properties of the final moulded product. The situation is comparable to that in rubbers, where the number and types of cross-links between the chains are very relevant to mechanical and ageing properties. For instance, long-term creep properties under tension and the resistance to impact are considerably affected by the amount of crystallinity, and on the orientation of the crystallites. Cracking under stress, particularly in the presence of certain solvents or detergent liquids is similarly affected. If anti-oxidant has to be added to the plastic to discourage the long-term effect of the weather, this tends to get segregated in the non-crystalline regions. The service life of the product can therefore be markedly affected by these subtle changes. A striking example of the effect of molecular orientation is seen in polypropylene, which like polyethylene, crystallises readily; thin sections of polypropylene can perform as permanent hinges, and will resist continuous flexing for very long periods, if the molecules are correctly oriented across the line of bending.

Other examples of the importance of texture in polymers can be given, at at various levels. In nylon and polyester fibres it is necessary to develop the correct crystallisation and crystallite orientation by drawing and annealing to get the desired tensile modulus and elongation properties. Even in polymers of low crystallisation tendency such as poly(vinyl chloride) (PVC), control of textures is important for practical reasons. PVC shows considerable dipolar interaction because of the presence of C–Cl bonds, and this

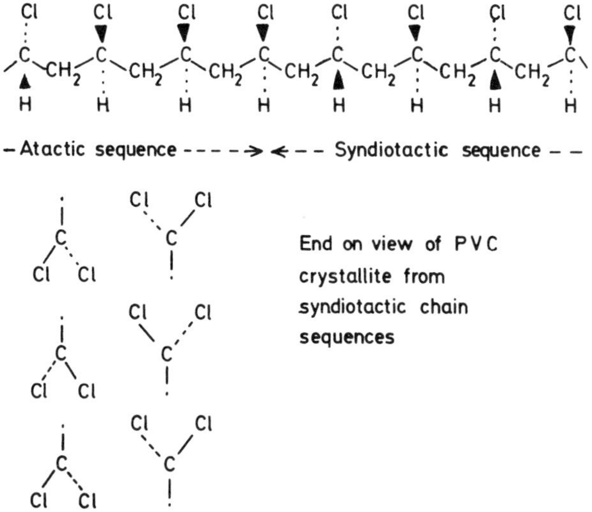

— Atactic sequence ----→ ←--- Syndiotactic sequence --

End on view of PVC
crystallite from
syndiotactic chain
sequences

Figure 2. PVC texture. Diagrammatic illustration of crystallite formation.

can be much reinforced if regular syndiotactic placement of monomer units units along the chain can be arranged; it seems that the syndiotactic structure packs more readily into a crystalline lattice than does an isotactic structure (Figure 2). Ordinary PVC, polymerised in a suspension of droplets in water at about 50°C, has a rather random structure, but there are enough crystallisable syndiotactic chain segments to form very small ordered regions. Once formed, these are relatively stable, and have a melting point of about about 170°C. Their presence may explain the difficulty of producing completely homogeneous rigid PVC extrudates and mouldings which is sometimes experienced (microscopy may show traces of the original PVC particles); and also the slow, but nearly complete, recovery from strain which takes place in some plasticised PVC's under conditions where visous flow and relaxation would rather be expected. Certainly, PVC's in which a relatively high degree of regular syndiotacticity has been developed (*e.g.* by polymerisation in urea channel complexes) are difficult to process by normal methods, seemingly because of the presence of these high-melting crystallites.

Another practically important aspect of PVC morphology may be mentioned, which is ultimately related to chain structure. During polymerisation of vinyl chloride in bulk (or in the monomer droplets during suspension polymerisation) the first-formed chains come out of solution as very small clusters, which aggregate with others to form the microscopic particles of the commercial PVC powders. By careful control of manufacturing conditions, which have been learnt by long experience, the particle size distribution, bulk density, and pososity of the product can be regulated; this control of particle morphology is vital to ensure satisfactory handling during moulding operations. For instance, rigid PVC pipes are made by extruding an intimate mixture ('compound') of PVC, stabiliser, pigment, and toughening additives. The compound can be pre-pelleted and fed to the extruder, but it can be more economic to eliminate this operation and to feed the mixed powders to the bigger machines, which then both compound and extrude. This only works properly if particle size and bulk density are suited to the design of the machine hopper and screw. Particle porosity can also be important, particularly in the plasticisation of PVC for the manufacture of flexible articles, sheet, or cable insulation, when the rate of imbibition of the liquid plasticiser, and the rate of gelation of the product must be under strict control. It can therefore be seen that, as in wood or steel, texture in solid polymers is most important, and will influence materially the mechanical behaviour of products fabricated from them.

Multiphase Polymers

So far, only chemically simple polymers mainly used in the plastics industry, have been mentioned. It has been shown that even in these the texture of the solid polymer can be very far from simple. Long before the details of this kind of complexity had been investigated, technologists had deliberately introduced artificial textures into polymers for good practical reasons, not only by such methods as stretching and drawing, as with fibres, but also by adding particulate or fibrous fillers. In effect, microscopic textures of all kinds can be produced by mixing polymers with fillers. Thus, a rubber tyre ought perhaps to be called a carbon black tyre, because the mechanical strength and abrasion resistance of rubber are vastly increased by the correct

incorporation of a considerable proportion of this material. Some phenol-
and urea-formaldehyde mouldings have been described facetiously as 'wood
flour glued together'; glass- and carbon fibre-reinforced polymers merit
as analogous description, but it must be said that in all these cases the 'glue'
is a vital, integral part of the very useful structures made from such materials.
The use of reinforcing fillers in plastics is of particular interest now, at a
time of scarcity and high prices (Plate 10). High-impact, or toughened,
polystyrene (TPS) and acrylonitrile–butadiene–styrene (ABS) resins owe
most of their toughness to the presence in them of a very large number of
small rubber spheres, which have the effect of increasing the internal absorption of strain energy at the tip of actual or incipient cracks. High-tenacity
polyamide fibres have a marked reinforcing effect on polypropylene.
Electrically conducting paint films, plastics and fibres can be made by suitably
incorporating into them conductive carbon particles; 'paint-on' electric
radiators and electrically heated carpets and clothing are possibilities for the
future!

The continuing interest in research on polymer textures is likely to result
in further improvements in the mechanical performance of articles fabricated
from polymers, since in this field it is often possible to carry over the more
academic results and theories into useful practice.

Art or Science?

This is a good point at which to ask how far academic research, as distinct
from industrial development, has contributed to our proficiency with
polymers. This is a difficult question to answer objectively and in any case
universities are not equipped materially or psychologically to do industrial
research. Langrish[2] has presented evidence, based on literature citations,
that, at least up to 1967, work in British universities on a number of topics,
including some on polymers, was of little industrial interest; in fact a small
prize, so far apparently unclaimed, was offered for a demonstration of a case
where real industrial benefit had resulted from such work. While the criteria
used by Langrish are of arguable diagnostic value, experience shows that the
great majority of advances and innovations in the polymer field have indeed
come from industrial (or industrially inspired) R. & D. efforts—for example,
emulsion polymerisation, redox catalysis, melt spinning, vulcanisation of
rubbers, control of molecular weight by transfer agents. All three industrial
methods for polymerising ethylene were developed essentially by the
intelligent systematic investigation of an accidental observation. There
are notable exceptions to this, and the names of K. Ziegler, G. Natta, and

M. Szwarc need only be mentioned; there is no doubt that some sections of the polymer industry have been revolutionised by the work of these scientists, from whom we have learnt to control the steric structure of polymer chains, and how to produce block polymers. Incidentally, the ability to carry over the 'aseptic' techniques used in this kind of chemistry, when a few parts per million of oxygen or water may ruin a reaction, to the 10,000 ton scale, is a remarkable tribute to the chemical engineers.

As is commonly the case with materials, the pattern of polymer development has usually been as follows. Industrial development produces innovations; aspects of these innovations are picked up by the academic world and researched, producing mechanisms, explanations and theories; these latter are absorbed by industrial development people whose experimentation then becomes more logical and efficient by being based on a developing body of reliable facts, concepts and quantitative relationships. As with other materials developments, the design of industrial polymers continues to be a mixture of science and art, seemingly as recognised by the Patents Acts! Polymers have become academically so respectable, that it is possible to take first and higher degrees in polymer science and technology. However, too often the scientist and the technologist seem to go their separate ways. Many of those who make and use polymer products are excellent practical technicians who are not very interested in polymer science, as they do not think it can help them; while some polymer researchers are much more interested in academic niceties than in utility. This is a pity, and it does not make for real proficiency, because either side could learn many relevant things from the other.

Some of the control methods for polymers now available are shown in in Table 5, with brief indications of how they are used; the original source of the idea which finally led to practical use of the method is also mentioned. A further point should not be forgotten; most polymers cannot actually be used without the assistance of antioxidants, stabilisers, plasticisers, dispersing agents, pigments, fillers and processing aids of all sorts. At a guess, the amount of work done in this practically important field may well be greater than that on the polymers themselves; most of it has been carried out in industry, although notable academic advances have been made in some areas, such as stabilisation and anti-oxidant action.

Finally, the amount of work put into the design and construction of the very large and costly polymer production plants by chemical, mechanical, and other engineers must be mentioned; scale-up methods have been developed, often by trial and error and bold experimentation, and com-

162 C. E. HOLLIS

Table 5: Control of Structure

Characteristic controlled	Examples
Chain composition	Random copolymers with uniform overall composition —from most vinyl and diene monomers Regular alternating copolymers—butadiene/acrylonitrile rubbers* Block copolymers—SB thermoplastic rubbers* —elastic polyurethanes —tough polypropylene* —polyester fibres —silicone/styrene blocks* —olefine oxide copolymers Graft copolymers—ABS, MBS and similar plastics —rubber/methyl methacrylate
Chain stereochemistry	Tactic polyolefins* Tactic polybutadienes, polyisoprenes* Tactic polymers of polar vinyl monomers*
Average molecular weight	Practically all commercial plastics, rubbers, and fibres; by control of chain termination
Molecular weight distribution	In many polymers—*e.g.*, polyethylenes or polystyrenes for specific uses
Type of cross-linking	In rubbers, for property control; polyurethanes, polyesters; paint resins
Particle size distribution	In polymer emulsions, latices; In TPS, ABS, *etc.* for toughness control; for processability, *e.g.* in poly-ethylenes, PVC
Particle shape and porosity	As for particle size distribution, *e.g.* in PVC

*Commercial development inspired by academic discovery
Key: A = acrylonitrile B = butadiene S = styrene M = methacrylate esters
TPS = toughened (or high-impact) polystyrene

merical polymer production has become more and more efficient as the use of continuous processes, very large reactors, and complex instrumentation and control methods have been brought in. Of necessity the industry has become highly capital-intensive, and the problems of developing new materials are very serious. Even before the onset of the present difficulties over energy and raw materials, few firms were able to tackle the introduction of completely new polymers because of the cost of the long and anxious development period virtually certain to precede the acquisition of a reasonable market. It seems likely that what chemical expertise has been gained will in the immediate future be employed in improving and varying the properties of existing polymers based on readily available raw materials, rather than in developing new ones.

Very few completely new commercial polymers have appeared in the last few years. In 1969, Willbourn[3] listed 25 plastics which had been introduced since 1964, mostly based on fairly readily available building blocks. The most exotic of these were the polyimides, polybenzimidazoles, polyarylene oxides, and polyphenylene sulphide. Some of these have since gone into sizeable commercial production[4] and are being used where their specific properties warrant their cost, that is, in special-purpose articles where raw material price is not particularly relevant, and performance is the main criterion. High-tenacity, heat-resisting fibres,[5] and new types of rubber, e.g. polypentenamer[6] have also appeared, but new commercial plastics are hard to find.

It would be unwise to be too dogmatic about the commercial emergence of new polymers; a good many have turned up during research on structures resistant to very high temperatures or to the exigencies of space flight; many will never get past the laboratory door and the few that do will remain expensive specialities. However, we must not forget the history of polytetrafluorethylene, discovered accidentally, which started as a costly material devoted to classified uses, but which now graces most of our frying pans.

Before turning to other aspects of polymer technology it should be said that many developments are taking place in different areas. Interesting new things can be found in the pages of the polymer journals, and doubtless some of them will evolve into something more useful than mere theses. A selection of polymers which have been described recently is given in Table 6.

Fabricated Products

Chemistry, as such, never produced so much as a moulded ashtray. Polymers may be designed in the laboratory and produced by the ton in a costly factory, but as materials they are of no particular use until someone has converted them into carpets, conveyor belting, or roofs for sports stadia. This brings us into the province of the engineer, the physicist, and the materials scientist.

As mentioned in Table 1, an important characteristic of polymers is that they can be accurately shaped or moulded and do not need complex and expensive machining operations to be performed on them. (They need complex machines to process them, but this is a common factor in the fabrication of any material.) Polymers also exhibit many mechanical and other properties which are of great practical utility; for instance they can combine strength and lightness, at a cost which has been low enough to ensure their wide application.

As to shaping, polymers can be pushed through spinnerets or dies continuously, to give fibres, profiles, pipes, or films; they can be compressed

Table 6:

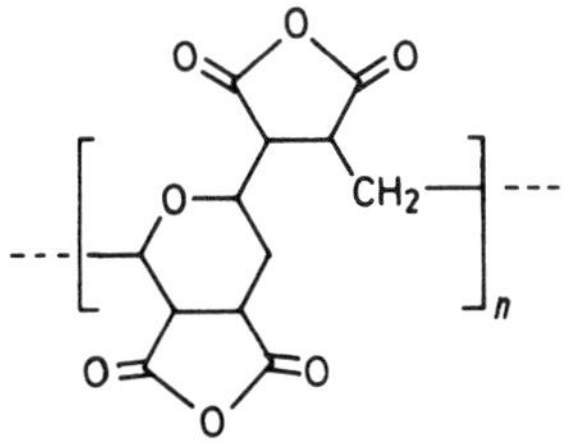	Copolymer for anti-tumour studies (Breslow, Edwards, Newburg: *Nature*, 1973, **246**, 160)
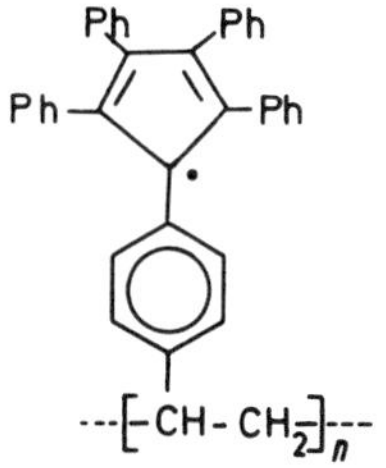	Polyradical (Braun: *Pure and Appl. Chem.*, 1972, **30**, 49)
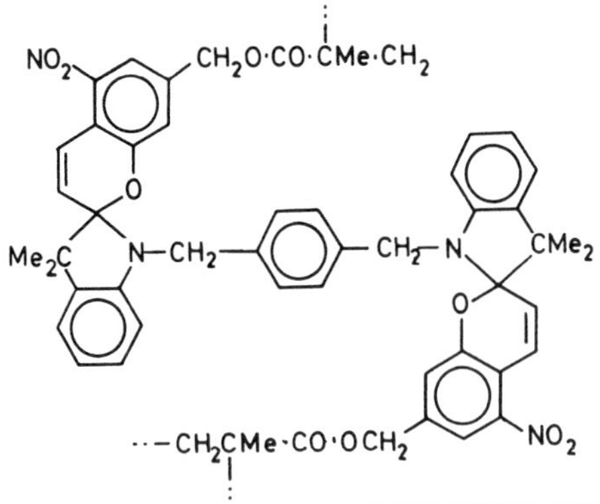	Photochromic/photoelastic polymer (Smets: Paper to S.C.I. (1973))
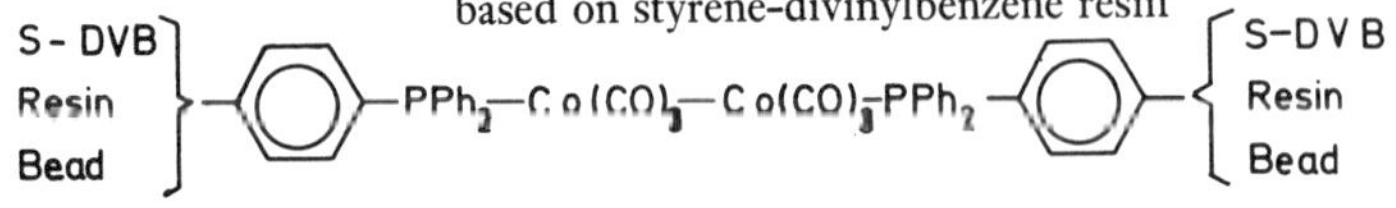	Hydroformylation catalyst for olefins based on styrene-divinylbenzene resin

Table 6: *continued*

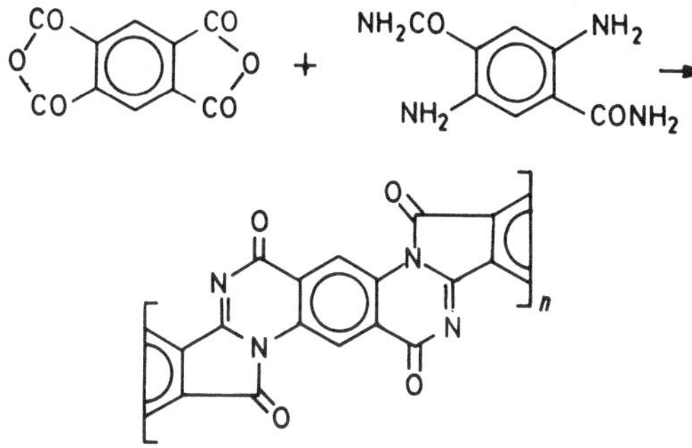

'Ladder' polymer stable (TGA) at 450°C Rabilloud *et al.*, *Makromol. Chem.*, 1967, **108**, 18.

ENKATHERM flame-resistant fibre developed by AKZO (Frank: Plastics Institute Research Meeting, 17th April, 1973)

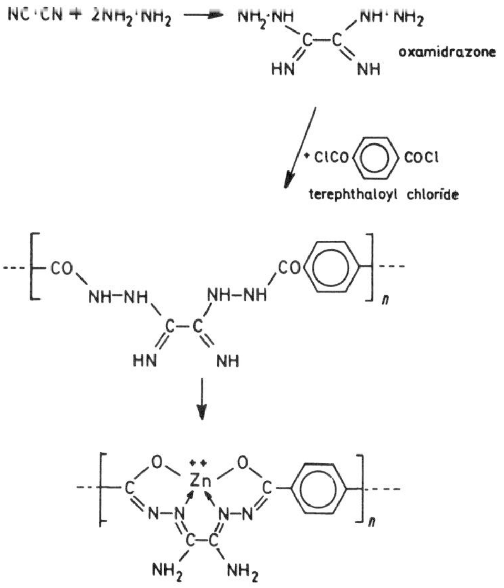

$NC \cdot CN + 2NH_2 \cdot NH_2 \longrightarrow$ oxamidrazone

terephthaloyl chloride

in moulds on a repetitive basis to give articles with complex shapes and accurate dimensions; they can also be cast, sintered, calendered, or converted into rigid or flexible foams. They can be combined in many ways with other materials to give composites. Most of us are reasonably familiar with products made in these ways, but the actual methods and machines may not be so well known, nor the problems presented to the engineer. These will be reviewed briefly because they are very important in the context of proficiency; they are mainly concerned with (a) how best to deal with polymers in the moudable condition and (b) how to ensure that the final product has the properties required of it.

Here we can only look at some elementary points which are common to most moulding processes, and attention will be focused on thermoplastic moulding materials, bearing in mind that processes for moulding thermosetting plastics or for vulcanising rubber products (while subject to similar constraints) will present their own problems because chemical reactions leading to infusibility take place during the cycle.

Moulding thermoplastics involves in very general terms the operations shown in Table 7. Melting is typically carried out in an electrically- or oil-heated vessel, which may be equipped with a screw, as in an extruder, and which conveys the material via melting and pumping zones through a die. With fibres, a melting chamber, gear pump, and spinneret provide similar functions. Apart from the problems associated with poor heat conduction and avoidance of thermal or oxidative degradation, the consistent control of visco-elastic behaviour in the polymer being moulded is of paramount importance. There is not space here to dwell on the complications caused by pigments, fillers, etc., which are often incorporated in polymer mixes.

Polymers have high melt viscosities which makes it difficult to force them through dies and into moulds; they have elastic memories because of the tendency for macromolecules, stretched and oriented in shear gradients, to relax when moulding constraints are removed. Difficulties associated with this kind of behaviour become more marked the higher the molecular weight (chain length) of the polymer being moulded; but on the other hand mechanical properties of the end product usually improve at higher molecular weights, so the moulder has to compromise. When the shaped polymer is cooling, some molecular relaxation will take place, as the molecules curl up, depending on how fast cooling takes place; the cold product will almost certainly be inhomogeneous (because relaxation and hence the residual molecular orientation will not be uniform) but will have in-built fluctuations of density, and associated internal tensions.

Table 7: Moulding Thermoplastic Polymers

Operation	Method	Problems
Convert granules or powder to molten or 'plastic' state	Heat ± mechanical shear	Poor heat conduction Thermal instability Oxidative instability
Force melt through die or into mould	Pressure, mechanical shear	High, non-Newtonian viscosity, visco-elastic effects—e.g., molecular orientation, elastic memory, die swell, melt-fracture
Solidify under conditions giving desired shape	Cool, ± pressure. Films and fibres mechanically extended (drawn) at this stage	Visco-elastic effects—e.g., molecular relaxation, crystallisation; causing internal density fluctuations and stresses

In the case of polymers such as the polyolefins and nylons, crystallisation will take place at the cooling stage, and must be controlled; this is particularly important in the case of fibres, in which the required texture is developed by various stretching and annealing processes.

The ultimate mechanical properties of the moulded article will be affected by these internal inhomogeneities in complex ways; thus tensile moduli or impact resistance may well depend on the direction in which stress is applied. It will be appreciated that to carry out a proficient moulding operation, one needs to consider very many factors, full information on which is difficult or impossible to get. Even if full data on any one factor were available, it would not necessarily be possible to relate the theoretical situation to a practical, controllable process, or to some readily measured property of the moulded product. Although much physical research has been done on many aspects of polymer processing—melting, flow of non-Newtonian liquids, visco-elastic melts, heat transfer, residual molecular orientation, crystallisation—practical situations are likely to be so complicated that theoretical guidance towards the solution of moulding problems can only be given in general terms.

Nevertheless, progress is being made; for instance the flow of molten polymer inside a screw extruder is now a fairly well-known process which can be treated theoretically with enough success to enable optimum screw profiles for specific purposes to be calculated by computer methods.[7] The granule-melting stage is not yet so fully analysed.[8] Theoretical calculation of die profiles to produce particular extrudate cross sections is possible in certain cases, although most die design has still largely to be based on experience plus trial and error. In sum, an increasing amount of fundamental information is now at the disposal of machine design engineers.

The best conditions for a given polymer shaping operation must usually be found by experiment and experience; once conditions are set up, control of machine variables can then be maintained. This is facilitated on modern machines by increasingly sophisticated instrumentation, which permits close monitoring and control of all mechanical movements, pressures, pressure gradients, temperature cycles, and so forth (Plate 11). Pre-set automatic control is commonly used, and computerised feed-back control is now being investigated.[9] The capital cost of big machines is already high, and the heavy instrumentation necessary for such control methods may well increase this by 10—15%; apart from this, maintenance of instrumental and machine accuracy needs highly skilled attention, since otherwise the advantages of close control can easily be wasted.

There has been no lack of research into many of the problems associated with these processes, and many papers have been published by physical chemists, engineers, and physicists, in many countries. In the U.K., the Science Research Council supports several projects in the universities. Much information of use to machine designers and processers has resulted, but there seem to be serious barriers to its penetration into all the places where it would be of benefit. It is partly a matter of interpretation, and partly one of availability of skilled people. The theoretical analysis of moulding problems soon leads to difficult concepts and complex mathematics, and the translation of the results into a language which is intelligible and useful to the practical engineer who has to run a moulding shop is not easy. Not too many engineers have the opportunity of studying the materials science of polymers. However, there seems no doubt that this is an area in which the technology would benefit by becoming more science-based; this has recently been stressed by the N.E.D.O. Plastics Working Party,[10] which has called for greater technical attention to design and processing in the U.K. industry.

Materials and Design

We have now come full circle, and must return briefly to the mechanical properties of polymers. When it is necessary to design and produce an article using a polymer, whether a tyre, a fabric, or an office chair, one of the first things to decide is what polymer to use. This may sound a simple matter, but it must be remembered that physical information on polymers is not yet systematised to the extent that it is for, say, steel; it may be quite difficult for a non-specialist engineer to find, let alone interpret, because although much is published, many of the data (especially for plastics) are not in a form suitable for design purposes. Selection raises questions about what the product is intended to do, how it is to be fabricated, and how much it will cost. While increasingly such questions are expertly analysed before a decision is made, by most manufacturers who hope to survive, there are still far too many cases where insufficient attention is given to finding the right material. The answer to some selection problems is of course fairly obvious; thus, there would not be much point in basing an oil-seal for a motor-car bearing on a rubber which swells heavily when exposed to lubricating oil, so that the choice would probably be for an acrylonitrile-butadiene rubber. Thereafter, all sorts of detailed questions about specifications, dimensional tolerances, mechanical properties, shape, compound formulation, moulding schedules, and price would have to be answered, mainly on the basis of previous experience. The ultimate sanction with components and articles for industrial use is an economic one; if the item does not function properly as a result of bad selection of material, faulty design, or poor moulding there will be trouble, which is rapidly reflected back to the producer. This is wasteful, but to some extent self-correcting in the long run, as the incompetent will finally go out of business. Whether this is sufficiently true of the large number of articles based on polymers which are sold direct to the general public is quite another matter. It is unlikely that in future we shall be able to afford to use expensive materials and machines in producing ill-designed and badly moulded plastic articles, some of which are still offered for sale.

Selection, then, should be based firmly on the intrinsic properties of the synthetic material from the design stage onwards, and not just on their possibilities as low cost substitutes for something else. The attempt to reproduce identically a particular article or component previously made of a 'traditional' material (e.g. of metal) will in most cases lead to trouble or even disaster. The range of properties which must be considered when a plastic article is being designed and produced in quite extensive; some factors will be advantageous to such use, others will not, and often a balance must be struck. Table 8 shows some of the factors which need consideration.

Performance under stress is naturally very important in the case of load-bearing components; even in less critical uses, the long-term effects of apparently minor stresses can be annoying or worse. Because of their macromolecular nature, as has been explained, polymers in the solid state

Table 8: Material Selection—Some Properties to Consider

Density
Thermal properties—
    expansion, conductivity, specific heat, *etc.*
Mechanical properties—
    moduli, ultimate tensile strength, *etc.*
    long-term stress: time/temp./load effects, creep, fracture
    short-term stress: impact fracture, notch sensitivity
    T_m, T_g, brittle transition, *etc.*
Electrical properties—
    resistivity, dielectric strength, tracking,
    dielectric behaviour at various frequencies
Resistance to environment, *etc.*—
    O_2, moisture, light, weathering, chemicals, solvents; barrier properties against gases
    and vapours
Optical properties—
    clarity, colour, finish
Flammability—
    resistance to ignition, flame, fire performance
Methods of fabrication—
    moulding, *etc.*, reinforcement, foaming, laminating
Cost—
    raw material, per unit wt., unit-volume, unit property of interest; conversion costs;
    cost per finished article
Other—
    possibility of recycle or disposal

---

show complex time- and temperature-dependent behaviour when subjected
to stress. The visco-elasticity shown in the liquid state is carried over into
the rubbery, glassy, or semi-crystalline solid condition, although the time
scale of the effects is very different in the various regimes. One important
practical effect is that materials like PVC or the polyolefins may show
appreciable change in dimension (creep) when subjected to continuous
loading at ordinary temperatures. The creep may relax if the load is removed. This effect must be taken into account when designing articles which
will be subjected to stress during their lifetime. Creep is in general non-
linear with respect both to loading and temperature, and if it goes on long
enough it may in some cases result in rupture of the material under loads
which are nominally much less than the ultimate breaking stress shown in
short-term tensile or shear tests. In some cases also (*e.g.* nylons) creep is
affected by the ambient humidity.

Creep testing needs meticulous care and takes time.[11] A complete theoretical analysis of creep phenomena is not really practicable (partly because

of the non-linearity mentioned), but a great quantity of experimentally determined data is available on many plastics and fibres for use by mechanical engineers and designers. In the case of composites such as glass-reinforced plastics, the presence of two components complicates experimental and theoretical treatment of this subject. Suitable methods for using creep data in standard design formulae, are available,[12] and the general position can be summed up by saying that design of load-bearing components in plastics is fairly well advanced; however, the massive data which are readily accessible in the case of steels and other traditional materials cannot yet be matched; and many design engineers are still rather shy of polymers. Continued experience with structures, from GRP ships' hulls to gears for watches, will slowly improve the situation.

A great deal of attention to this question of 'rational' design has been given recently,[13—15] and a high degree of success in the intelligent use of plastics has often been achieved. As an illustration of some of the complexities involved, we may consider briefly the use of high-density polyethylene (HDPE) for the blow moulding of large drums and similar containers. Some of the chemical, physical, and mechanical complications with this material have already been mentioned as a background. There is no space to go into details of materials selection here. The main competitors on overall technical and economic grounds would be low-density and high-density polyethylenes and propylene copolymers. LDPE is similar in many ways to HDPE, but has a much more branched-chain structure, which hinders crystallite formation; its density and rigidity are lower because there is more of the flexible, amorphous phase present so that to obtain equivalent rigidity, thicker sections would be necessary than with HDPE. Polypropylene is harder, more rigid and with a higher melting point than HDPE.

Table 9 gives some of the qualitative design criteria for the drums. It will be seen that a good deal of information must be obtained on how the product will be used, before a rigid drum which will not distort unacceptably during stacked storage can be designed. Blow moulding is a fairly complicated operation (Figure 3) (Plate 12), and Tables 9 and 10 show some of the practical problems which have to be solved in connection with machine and moulding powder; both have to be fitted to their purpose. This very condensed account of a typical blow moulding operation could be elaborated into a separate paper, which would not be of direct interest to the average reader; it is included merely to show that the chemistry of materials is only part of the story. A product must be properly designed and fabricated correctly

Table 9: HDPE Drums—Some Design Points

Capacity—What size? Returnable? Shape, for handling, packaging, storing, stacking? Colour?

Moulding Method—Single, multiple step? Type of machine? Size, capacity? Method of feed (granule, powder)? Take-off, finishing operations? Scrap recycle?

Mechanical—What loading, stacking, use conditions? Highest, lowest, mean temperature? Other factors? Bursting, impact performance? Are strength, creep, impact data available?

Corrosion—What contents, cleaning liquids? Risk of environmental stress cracking? Exposure to weather?

Testing—Methods, specifications, test rigs available?

Cost—Comparison with equivalent metal drum? Recycle, cleaning advantage? Recovery of plastic for reuse? Break-even production?

---

in the right materials to meet the real conditions under which it will be used, and unless *all* the links in this chain of responsibility are sound, unsatisfactory performance or even failure may result. If failure does occur, it is quite likely that the material alone will be blamed, probably unjustly. What the customer finally gets depends as much on engineers as on chemists, and it is important to make sure that full information on materials and design methods is easily at hand for engineers.

It seems altogether too Utopian to believe that the materials technology of polymers, with its wide implications, will ever be fully scientific in the sense that in the millennium nice, tidy theories will prescribe by computer exactly what should be done during all stages of producing a floor tile. Nevertheless it is most desirable that polymer chemists and physicists should appreciate the needs of polymer engineers more than they do at present, and that all concerned should at least know where to go for good advice.

Proficiency and Economic Changes

So far, only simple technical aspects of the proficient use of polymers have been mentioned. Our chemistry and physics are not too bad, although there are still many questions to be answered, but our design and fabrication methods could be improved if we tried hard enough and directed our research into more practical channels. Ultimately, however, we come up against elusive economic and sociological problems. What ought we to be doing with polymers now? Have we been prodigal rather than proficient?

It seems likely that it is technically possible to solve our energy problems in due course, politicians permitting. The raw material problems solution depends on whether or not it is possible to devise world economic systems

PLATE 7 Acetal copolymer watch movement (Tissot), from Hoechst Chemicals moulding materials. *Courtesy 'Plastics & Rubber Weekly'*

PLATE 8 Intricate mouldings for Polaroid Swinger camera. *Courtesy 'Plastics & Rubber Weekly'*

PLATE 9   Submarine cable connection in PVC–Nitrile–butadiene rubber compound. *BP Chemicals International Ltd*

PLATE 10   Hull of 'HMS Wilton' minesweeper in glass reinforced polyester

PLATE 11   Large injection moulding machine. Bone-Craven, 450 oz, for Chloride-Lorival. *Courtesy PRW*

PLATE 12   Experimental blow-moulding of high density polyethylene drum, in 'Rigidex H'. *BP Chemicals International Ltd*

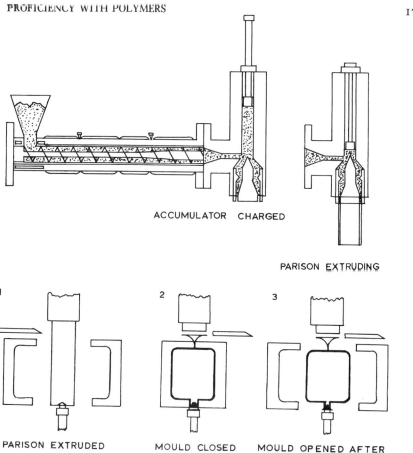

ACCUMULATOR CHARGED

PARISON EXTRUDING

1 2 3

PARISON EXTRUDED   MOULD CLOSED   MOULD OPENED AFTER
MOULD OPEN   PARISON BLOWN   COOLING PERIOD

Figure 3. Ram accumulator system for the blow moulding of polyethylene. [Diagram courtesy of B.P. Chemicals International Ltd.]

which are stable under near-equilibrium conditions, instead of operating in growth–collapse cycles. In spite of the lessons of history and biology, hope of achieving this still seems to spring eternal in the human breast, but this is not the place to discuss such matters; we are likely to be more interested in the next five years. It is not practical to make much change in the source of polymer intermediates, chiefly oil, in the short term, although doubtless the use of natural gas and coal will increase. Increased production

Table 10: Problems in Blow-moulding

	Material and Machine Requirements to be met	
Operation	Material	Eng. design and control
Feed powder to screw	Bulk density Particle size distribution	Hopper shape/feed to prevent clogging Screw conformation to accept powder Heat input/output
Melt powder in screw	Melting behaviour	Screw conformation for melting/mixing Heat input/output
Transfer melt to accumulator	Stabilisation against thermal/oxidative degradation	Screw conformation for pumping Heat input/output
Melt through die to form parison	MW and MWD for good flow, min. relaxation (low MW), max. melt strength to avoid sag (high MW). (compromise)	Die for optimum parison profile Temp./pressure control
Blow into mould	Good melt flow	Complex mould and inflation device Temp./pressure control
Cool moulding	Crystallisation rate and reproducibility Nucleation	Temp./pressure control
General		Automatic cycling Instrumentation

MW = Molecular weight
MWD = Molecular weight distribution

of cellulose, lignin, wool, and rubber from agricultural sources cannot be organised in a short period. Hence, the main questions will be concerned with the price and availability of current materials. It is safe to predict that polymer prices will increase, probably relatively more than in the case of other materials, since there is much evidence that the polymer industry has been working within margins which are too narrow to allow for reasonable reinvestment. At the same time the present price differential between high performance and general-purpose polymers may well fall, as the scale of production of the former increases. These price changes ought to encourage the use of polymers in the manufacture of the more durable and essential products, at the expense of short-life and throw-away articles.

Most of us would accept that in very many cases polymers are being used sensibly and efficiently as major or minor components of many essential products; without them the present fabric of our society would come apart at the seams. On the other hand we recognise that the words 'sensible and efficient' do not apply to each and every use of polymeric materials, especially plastics. The demands rather more than the needs of a consumption-oriented society have provided the driving force for the expansion of the polymer industry, but the development of mass markets, and supermarket selling techniques, has not always gone hand in hand with quality; far too many products have appeared, in connection with which it would be complimentary to talk even of 'planned' obsolescence, 'unplanned' being nearer the mark. We all have our little lists of under-designed gadgets, styled for looks rather than function, which have aroused our ire. The many and complex attitudes of the public, to plastics in particular, have been brilliantly analysed by Glanvill;[16] ultimately what the public demands from polymers will settle what is produced. It is therefore important to ensure that these demands are realistically based. This is partly a matter of information and education, which requires time, and considerable efforts are being made in this direction by industry and the professional bodies; thus, assistance is being provided in the schools on many aspects of polymers as materials, while at the consumer end the use of labelling schemes (such as that giving standard instructions for washing and cleaning fabrics) is increasing. Incidentally, in these days of university 'Science Greats' courses, what could be a more useful subject to the educated citizen than materials science?

The present crisis over materials supply and prices has given a severe check to the polymer industries. It will result in many problems of substitution, conservation, and recycle. Plenty of technical knowhow is available to solve these problems, given a rational approach. Let us hope that as an outcome we can in fact increase our proficiency with polymers.

Bibliography

[1] E. G. Couzens, V. E. Yarsley: 'Plastics in the Modern World', Penguin Books, Harmondsworth, 1968, p. 350, attribute this phrase to Ellis. I have not been able to locate it in Ellis' monumental 'Chemistry of Synthetic Resins', 2 vols, Reinhold, New York, 1935.
[2] J. Langrish: Chem. in Britain, 1972, 8, 330.
[3] A. H. Willbourn: Plastics and Polymers, 1969, 37, 421.

4 (a) 'Modern Plastics Encyclopedia', McGraw Hill, New York, annually. (b) *Reports Prog. Appl. Chem.*, Society of Chemical Industry, London, annually. Both are good sources of information.

5 R. F. Wolf: *Rubber J.*, 1972, **154** (Feb), 23.

6 A. J. Amass: *Brit. Polymer J.*, 1972, **4**, 327.

7 (a) J. R. A. Pearson: 'Mechanical Principals of Polymer Melt Processing', Pergamon Press, Oxford, 1968. (b) Z. Tadmor, I. Klein: 'Engineering Principles of Plasticating Extrusion', van Nostrand Reinhold, New York, 1970.

8 A. Berlis, E. Broyer, C. Mund, Z. Tadmor: *Plastics and Polymers*, 1973, **41**, 145.

9 J. van Leeuwen, Paper 3; H. Keller, Paper 10; Conference on New Techniques in Extrusion and Injection Moulding, Plastics Institute, Manchester, 11th—12th April, 1973.

10 'The Plastics Industry and its Prospects', NEDO; HMSO, 1972.

11 S. Turner: 'Mechanical Testing of Plastics', Butterworth, London, 1973.

12 'Plastics and the Design Engineer', Plastics Institute, London, 1968.

13 P. C. Powell: 'Plastics for Industrial Designers', Plastics Institute, London, 1974.

14 R. M. Ogorkiewicz: 'Engineering Properties of Thermoplastics', Wiley, New York 1970.

15 P. C. Powell, S. Turner: *Plastics and Polymers*, 1971, **39**, 261.

16 A. B. Glanvill: *Plastics and Polymers*, 1973, **41**, 242.

Keeping it Clean

The river Rhine, it is well known,
Doth wash your city of Cologne;
But tell me, Nymphs, what power divine
Shall henceforth wash the river Rhine?

S. T. Coleridge

Clean the air! clean the sky! wash the wind! take stone from stone and wash them!

T. S. Eliot

The Biological Effects of Chemical Substances

by M. W. Holdgate; Central Unit on Environmental Pollution, Department of the Environment*

POLLUTION may be defined as 'the introduction by man into the environment of substances (or energy) liable to cause hazards to human health, harm to living resources or ecological systems, damage to amenity, or interference with legitimate uses of the environment'. This definition brings out the basic fact that in responding to pollution we are concerned with actual or potential adverse effects—damage to our own physiology, to the living resources on which we depend directly, or to the wider ecological systems of land and sea that play a vital part in the renewal of atmospheric oxygen, the elimination of many of our wastes, and the recycling of vital nutrients. In this paper I am concerned with the effects of chemicals injected into the biosphere by man (whether injected directly or indirectly *via* the physical environment), and the nature of our response to them.

Even the most apparently stable of living systems exhibits dynamic equilibrium. This is affected by physical and chemical variables at many levels. A pollutant in the environment may affect

(a) an ecosystem
(b) a population of a single species
(c) a single individual organism
(d) an organ or system within that organism
(e) a biochemical or cellular subsystem

or several of these simultaneously.

Hierarchical Effects of Pollution

Commonly, we consider pollution in terms of its effect upon whole individual organisms. It is evident that the responses of individuals are in fact the integral of a complex web of subsidiary effects upon biochemical systems,

* The views expressed in this paper are those of the author and not necessarily those of the Department of the Environment.

leading to changes in cellular physiology and behaviour which in turn affect the functioning of whole organs. A pollutant may often have a demonstrable effect at biochemical or cellular level which, because of the homeostatic machinery of living creatures, may not be manifest at the level of the whole individual at all. This does not mean that the individual is unaffected by the pollutant: indeed, it means that some of the biological capacity of that individual is being used up in the same way that the discharge of a pollutant into a lake uses some of its dilution capacity even though it does not reach the threshold at which ecological change ensues. Similarly, the responses of whole ecological systems are the integral of many interdependent individual responses. It is quite common for significant effects on individuals not to be manifested at all at the ecosystem level, where so many organisms of so many different species interact to produce a system that is in dynamic equilibrium and which displays considerable inertia.

These levels of effect are arranged hierarchically. The ecosystem level of response is the highest level and a "no–response threshold' may yet be maintained while there is a considerable pollution-induced mortality at the individual level, and even at the species level should the species exterminated from the ecosystem not be of major importance in determining the characteristics of the whole. For example, it has been established that exposure of lichen species on the trunks of English oaks to mean annual sulphur dioxide concentrations of 40—180 $\mu g/m^3$ can be correlated with their progressive extermination—and, no doubt, that of the small insects that feed on them. This has no effect, so far as we know, on the basic ecological stability of oak woods in the Midlands or South-East England, and, while it clearly leads to a degree of impoverishment compared with the natural situation, does not prevent the establishment of woodland National Nature Reserves which support an otherwise rich and varied wildlife. Similarly, it would be perfectly possible to contemplate a pollutant effect which exterminated some species from a freshwater or marine environment without affecting either the capacity of those waters to support a balanced and productive ecosystem or creating a system unacceptable to man.

In this situation pollutants can be considered to impose 'stress' on the ecological system just as other forms of human interference do. It seems probable that the very first effect is to increase the species diversity of the system, but thereafter rising levels of pollution are paralleled by reduction in diversity and also by a tendency to change from systems dominated by large, long-lived forms to those dominated by smaller species with shorter life cycles—as when woodland gives way to grasslands. In more technical language, the shift is toward systems in which the annual production is larger in proportion to the standing crop. This is not the place to go into details, but we do now know a good deal about how ecological systems respond to stress and this is valuable in predicting what new levels of pollution may do to patterns of vegetation and fauna.

At the population level, if one is concerned simply to have 'no response' in terms of population size, enhanced mortality is acceptable so long as this is not on so great a scale that it leads to population decline. That is, it is permissible in these terms to substitute pollution for other forms of mortality that would otherwise remove the surplus of offspring produced in most populations of living creatures. If the annual production of young seabirds, for example, is three times that required for recruitment into a stable breeding population and the norm is for the surplus two-thirds to die, in population terms it is immaterial whether this surplus is removed by competition for food, predation, or chemical pollution, so long as recruitment is sustained. The size of the population would be equally unaffected if the mortality among breeding adults were increased by pollution, so long as enough young survived to maintain recruitment at the higher rate which would then become necessary.

Indeed it seems very probable that during the early 1960s the heron population in England was in just this position in relation to certain organochlorine pesticides. We know that some adult herons were lethally poisoned and there is considerable evidence to suggest that sub-lethal residues affected egg-shell thickness and chick survival, resulting in a reduced output of young. Neither the adult mortality, nor the decrease in over-all breeding success, however, appear to have reached a level where they affected more than the annual surplus present in the population and so the total heron population of England and Wales of 4,500 pairs remained unchanged.

Again, a demand for 'no response' at the individual level permits organ, cellular, or biochemical changes, so long as mortality or a serious retardation of growth or distortion of behaviour does not ensue. The influence of factories emitting fluoride on cattle grazing in their vicinity has been regulated on such a basis: a 'no-effect' standard at the physiological level has not been demanded, but the exposure has been restricted in such a way that unacceptable effects are avoided. Rather similarly, some pasture grasses and crop plants in industrial areas of Britain may have their growth rates affected by atmospheric pollution, including pollution by SO_2, but this is regarded as acceptable so long as yields are maintained above a certain threshold, determined by economics. Even in man one sees this situation, although the threshold is usually set much more stringently on health grounds: *e.g.*, in our acceptance of blood lead levels which affect the activity of the enzyme δ-aminolaevulinic acid dehydratase, so long as these are not permitted to rise above around 36—40 μg/100 ml of blood, which is commonly taken as the point beyond which we become socially concerned. Finally, if we determine that we shall have a 'no-response' situation at the biochemical level, we are admitting of no detectable effect whatsoever and are faced with the most stringent of all demands for pollution control. Indeed, with an impossible situation, since some substances we release as pollutants are present naturally in our environment at levels which have a biochemical effect.

In social terms, we tolerate different levels of effect according to the nature of the target and its immediacy to the human situation. For example, it is commonly agreed that we should not accept significant effects at the organ level in man. We are indeed wary about the cumulative effects of lifelong exposure to substances which may only have minor demonstrable effects on human biochemistry, fearing that they may in some way progressively erode the body's homeostatic machinery and shorten life. We reject significant effects at the individual level—illness or impaired growth—in cattle, sheep, other farm stock, and domestic pets. We reject significant ecosystem effects in marine and freshwater ecosystems and in forests and wild vegetation, with their invertebrate and bacterial components. Sometimes, of course, the standard we set at one level affects organisms for which we would otherwise only be seeking lesser degrees of protection: standards set for mercury in fish as human food for example, may lead to curbs on discharges of that metal to the sea which from the point of view of protecting marine ecosystems would not be necessary. It does not follow that our judgement in these matters is always wise (although I believe that it is defensible) but it does follow that we should be aware of the logic of our actions.

Exposure and Effect

The effect on a target depends upon three groups of variables:

(a) the nature of the pollutant,
(b) the biological state of the target with which it is interacting, including the individual and temporal variation which is exhibited by individuals, organs and biochemical systems,
(c) the concentration of the pollutant and the time of exposure (not usually, incidentally, capable of being summarised by a simple arithmetical multiplication. Short-term exposures to very high concentrations need not

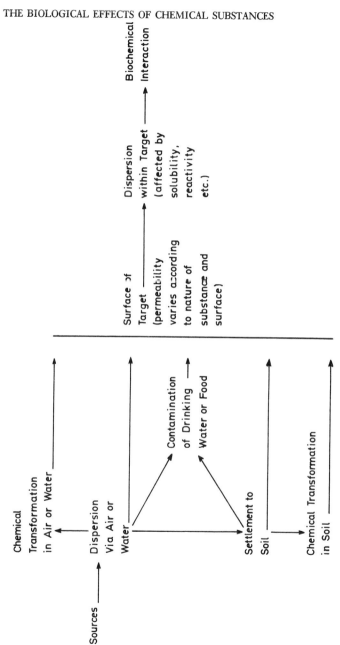

Figure 1. The pathways of pollutants from source to the point of effect.

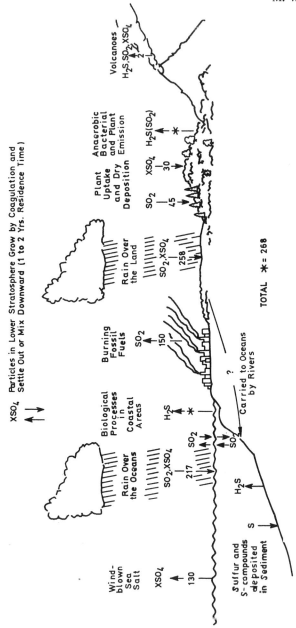

Figure 2. The pathway of sulphur dioxide in air (from W.W. Kellogg, R.D. Cudle, E.R. Allen, A.L. Lazarus and E.A. Martell, *Science*, 1972, 715, No. 4022, p.594.)

always be the equivalent of exposures to half that concentration for twice the time or one-tenth the concentration for ten times as long. Ozone for example, can affect sensitive plants in hourly exposures to around 10—20 parts per hundred million: twice the time at half the level may have much less effect). In addition, it is important to known the site of entry of the pollutant (*e.g.*, skin, lung, gut) and the site of action (nervous system, blood, kidney, liver, *etc*).

It is worth emphasising that exposure has to be measured *at the target*. The pathway from the source of pollution to the target is all important in determining exposure levels. Such a pathway may be represented as in Figure 1. It is evident that a whole series of physical and chemical properties and interactions influence this movement, and parameters such as diffusion coefficients, absorption properties in and from soil, solubility in water or other liquids, rates of breakdown by ultraviolet radiation, and reactivity with other components of air, water, or soil can have a major influence. This is well exemplified by the pathway of sulphur dioxide in air (Figure 2). Here transformation to SO_3^{2+} or SO_4^{2+}, washing out in rain as dilute sulphuric acid and/or combination with ammonia to form an ammonium sulphate haze ('Teeside mist') have an important influence on the actual nature of target exposure. Similarly, chlorofluoro-hydrocarbon compounds emitted to the environment as aerosol propellants apparently persist for a long time under ground-level conditions but are fairly rapidly degraded by ultraviolet radiation in the upper atmosphere: here physical mixing of layers of air will clearly be an important determinant of persistence. In predicting the scale of a possible pollution problem, therefore, any model we develop must take full account of the pathways, as well as the nature of the actual interaction of the pollutant and the target. It must also cater for individual variations in susceptibility and variations with time and circumstance.

Sometimes these variations can be very large. For example, concentrations of lead in air in Britain may reach 20—25 $\mu g/m^3$ in the centre of motorways by day and average 10—15 $\mu g/m^3$ over 24 hours. In ordinary busy streets, and around factories, 24 hour averages are around 1—3 $\mu g/m^3$. In the country, levels are much lower —below 0.1 $\mu g/m^3$. Human exposure inevitably varies as people move around. Human intake to the lungs likewise varies according to the volume of air we breathe in the day—and while the average for an adult man is around 15 m^3 this can be doubled by activity or halved by rest. Not all the lead particles passing the nose or mouth reach, or stay in the lungs: various calculations give retentions of 10—60% of the potential intake. Only a proportion of the lead retained is absorbed—maybe 50% on average, but with much individual variation. It you sum the cumulative effect of all these variables, you can readily see how there could be a thousand-fold variation in human lead absorption from the air—and the uncertainties in many figures cast doubts on the usefulness of such a broad-brush calculation at all.

Generally speaking, the effect of exposure to a pollutant is related to concentration, although the curve is unlikely to be linear. Moreover, the response to increasing concentration is frequently hierarchial in nature. That is, at the lowest concentrations there is a biochemical effect. With increasing concentration this becomes manifest in whole-cell or whole-organ changes, then in variations in individual behaviour. At higher concentrations still, individual mortality may ensure while greater exposure leads on to whole-population or whole-ecosystem changes.

We have partial documentation of many of these curves. Often the information results from laboratory tests, under conditions remote from those in the field, of single pollutants against single target species. This may be reasonable enough for exposures to pollutants with a specific action, like the effects of radiation on genetic systems, but how valid is it as an indicator of whole-population or whole-organism responses to chemical factors? Often the laboratory tests estimate doses that kill 50% of the exposed organisms (LD_{50}) in 24, 48, or 72 hours or over a longer period of days. Often, the dose concentrations used in such tests are relatively high—sometimes this is necessary to demonstrate a response in a convenient time. The test species used are often selected for convenience in the laboratory and basic (and sometimes sweeping) assumptions are made about the similarity of response other organisms may be expected to show.

But what of the situation in Nature, when we are dealing with mobile populations, forming part of a much more complex system, and able to move away from areas of high pollution? There can be no confidence in their extrapolation to the real world. Often, too, ecologically significant effects can be sub-lethal and will not show up at all in LD_{50} tests. For instance, seaweed growth rates can be related to degrees of marine pollution with conseqeunces for herbivorous marine animals; and fish physiology can be adversely affected by DDT levels below the directly lethal dose, making fish more likely to succumb to cold or to be unsuccessful on migration.

A good example of a complex and imperfect framework of knowledge is provided by sulphur dioxide in air—one of the commonest and most

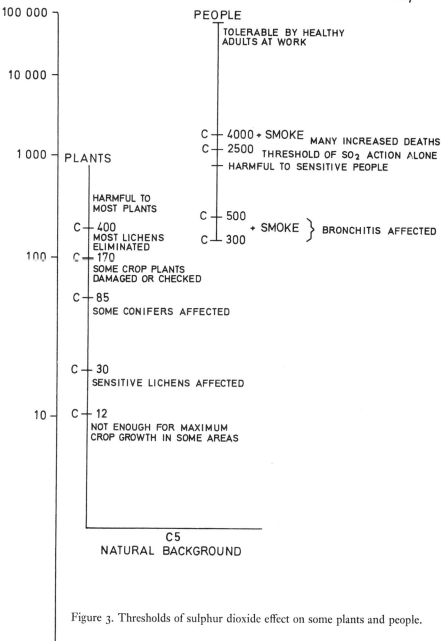

Figure 3. Thresholds of sulphur dioxide effect on some plants and people.

studied of atmospheric contaminants. As Figure 3 implies, quite a lot is known about the exposures at which various kinds of plant first show injury and then serious damage. Likewise, the onset of symptoms in the most sensitive people (bronchitics) is predictable. But in both cases allowance must be made for a very wide range of individual variation (in man there is probably a thousandfold range in sensitivity in an average population); furthermore, there is a great deal of variation in response according to the duration of the exposure, and at least in man, the levels of smoke accompanying the SO_2 exposure are critical. Thus, the 'threshold' level for SO_2, averaged over 24 hours, is 500 μg/m^3 when accompanied by 250 μg/m^3 smoke, whereas laboratory studies have failed to find significant impairment of human respiratory function by SO_2 alone at concentrations below 2500 μg/m^3 for comparable periods.

In this case we do have enough knowledge of dose/response relationships to conclude that in order to protect vulnerable people we should seek to keep daily average sulphur dioxide levels in air below 500 μg/m^3 and smoke below 250 μg/m^3 and these figures have been adopted as short-term objectives by a WHO Working Party. If we consider that the manifest sensitivity of lichens indicates potential biochemical disturbance which, while we cannot prove its parallel existence in man, ought in common prudence to be avoided, we would need to take an annual mean of around 40 μg/m^3 as our environmental quality goal—and see that peaks never greatly exceeded this figure. Although we have little information on other species, this would almost certainly also prevent any crop losses and protect most forms of life while it would not eliminate the beneficial action of air-borne SO_2 in correcting sulphur deficiency in some soils. The recently published national survey of smoke and sulphur dioxide indicates that 500 μg/m^3 is only exceeded in the south of England, excluding London, for a very few days in the year, whereas in London and in the midland and northern towns it is frequently exceeded. But in many rural areas the lower 'target' of an annual mean of 40 μg/m^3 is exceeded, so that it is not surprising that the British Lichenological Society has recorded considerable impoverishment of the lichen flora in the whole central midland zone of England. The question is whether this is paralleled by any significant agricultural losses (one report placed these at £40 million per annum in England, but with an enormous possible margin of error) and sufficiently serious to justify the major expenditure that would be required to reverse it (especially in the current energy situation).

Only rarely do we have even this degree of detailed information about exposure–effect relationships. More commonly all that we have is a broad,

descriptive correlation between pollution levels, generally assessed, and biological response: the Trent Biological Index is a well-known example for
fresh water. There is clearly a need for more research to determine rigorous
'criteria' on which our social judgement of the need for more (or less)
intensive effort to combat pollution must depend. Such work needs to take
into account the biochemical means by which the whole-organism responses
are mediated. We need a better understanding of individual variations in
response, and how variables such as age, nutritional state, or reproductive
condition of the individual interact with environmental chemical factors in
determining them.

The Interaction of Pollutants and Other Factors

This leads to another important generalisation; pollutants rarely operate
in isolation. I would like to give a case history that underlines this rather
well and illustrates the true complexity of the kind of situation that confronts
us. Between August and November 1969 about 12,000 seabirds died in the
Irish Sea. They were almost all of one species (*Uria aalge*, the Guillemot);
they were almost all adults, and nearly all of them came ashore after September gales in an emaciated condition. Investigation showed, however, that the
gales were not the primary cause, for while the peak in the number of strand-
ings coincided with the storms, some birds had begun to leave the water in
a distressed condition before the gales began and while gales of equal
severity had affected the whole western seaboard of Britain, the mortality
was concentrated in the Irish Sea north of a line from Holyhead to Dublin
and south of a line from Donegal to the Mull on Kintyre.

Histological examination showed that many of the birds had lesions in
the kidney and liver and changes in the heart and pericardium which paralleled experimental effects obtained by exposing birds to large concentrations
of polychlorinated biphenyls (PCBs). Chemical analysis subsequently showed
high levels of PCBs in liver and kidney, and to a lesser extent in brain. It
would be tempting, therefore, to frame the hypothesis that these birds died
of PCB poisoning. This, too, would be an oversimplification, since *healthy*
birds shot off the west coast of Scotland contained whole-body loads of
PCBs comparable with those in the victims of the disaster. The difference
was that in the healthy birds the PCBs were distributed in the body fat and
the levels in liver and kidney were low.

The conclusion of the study was that we were dealing with a multivariate
situation. The most plausible hypothesis was that something, possibly
climatic change, caused a reduction in the available food supply of these birds

in midsummer at a time when their reserves were fairly low after the breeding season and they were about to go into the moult, which imposes a further stress. The oceanographers were unable to detect any change in marine life, but since these birds feed only in the very top layer of the water, a long clear spell with bright sun might have led their food organisms to descend slightly deeper than normal, thereby shutting off the food supply but not creating changes detectable with the marine biologist's tow net. Be that as it may, the initiating cause of the mortality was very possibly an interruption of feeding. This would, of course, have led to the mobilisation of body fat, which would have brought the PCBs stored there into the blood stream and on their way to liver and kidneys. This in turn might well have had minor behavioural effects at the whole-organism level. It might have made the birds a little less efficient at feeding, thus exacerbating the situation, and more prone to exhaustion by storm, especially since their food reserves were already low; any further stress could only lead to the increasing flushing of PCBs into the blood. There would then be the recipe for a localised disaster in the area where the initial imbalance between food intake and food requirement had occurred, and the whole disaster would be attributable not directly to PCB poisoning, but to the impact of this pollutant within the system, tipping the scale between survival and death when the environment imposed external stress. I believe that this is likely to be a model commonly valid in the pollution area and that many pollutants must be looked on as operating in this way, changing the probabilities between health and disease in a situation depending on a great complex of variables.

The Prediction of Effects

In predicting what will happen under a variety of circumstances, it is clear that we must take account of the sensitivity of organisms to various levels of exposure to both individual significant pollutants and combinations of pollutants. We must also endeavour to predict the levels of exposure at these target organisms, and these in turn depend on the pathways of the pollutants and the factors determining their rate of input to and removal from the environment.

It is well known that all chemical substances vary in these respects. It is self-evident that the pollutants likely to attain significant concentrations in regions remote from their sources of emission are those released in large quantities, readily soluble in water or easily dispersed in air, having a rapid rate of diffusion and having a low rate of chemical or biological transformation into innocuous products or a low rate of settlement or sedimentation.

from air or water in which they are suspended, and perhaps ultimately exhibiting the phenomenon of concentration *via* food chains. Those of high persistence, but less ready dispersal, on the other hand, are clearly type-cast to create 'hot spots' about points of emission. In both cases, if we know about their inherent toxicity and the way by which that toxicity takes effect, we have the recipe for predicting the likely seriousness of a pollutant substance.

It would be useful were we able to go further than this, and predict from the molecular structure of a substance what its behaviour in the environment was likely to be. Sometimes this can be done to a degree, as with the probability of biodegradation of certain detergents. But a recent Royal Society discussion exposed the limitations of this approach. Even minor structural changes alter the behaviour of molecules to a surprising degree. For the foreseeable future we can expect to have to rely as the chemical industry now does on laboratory and field testing prior to the introduction of new substances or new formulations. What we can ensure however is that the mass of available information is more effectively retrieved, and this is the aim of the information referral system being developed for the United Nations, with the active participation of the UK Chemical Information Service.

The construction of meaningful numerical models of environmental pollution is extremely difficult. Not only is one confronted with the almost endless variation in biological response, but the curves relating emissions to concentrations in the environment at a particular point and time are influenced by a very large number of factors. We have at the present time some relatively crude models for the behaviour of pollutants in water, particularly rivers, which have the merit of being relatively confined and subject to predictable patterns of flow. Such models are being attempted for the sea, but are prone to great difficulty, while for air they have succeeded only at the most superficial level. Clearly, we need many more studies of pathways and effects, and these are the two key areas for research upon which must depend our social response in terms of judging the relative seriousness of a pollution problem and therefore the priority we must give to its control. For we must remind ourselves that we live in a world in which chemical factors are, and always have been, important in ecology and physiology. Many of the pollutants to which we are exposed are natural substances whose concentrations and distributions, rather than whose actions, have been modified by man. We cannot expect to eliminate those substances; indeed many of them are essential at low concentrations to our well-being.

We could not expect to take their concentrations back to pre-industrial levels even if we knew what these were. What we can do is seek to predict the points at which their concentrations become socially unacceptable and to work out the most efficient means, in economic and technological terms, of restricting those levels below that threshold. And some pollutants, *e.g.*, PCBs, are entirely unknown in Nature. Here also we must apply the principle of restricting their concentrations to the threshold of social unacceptability. On the basis of this knowledge, too, we can make a more sensible decision about the parameters we need to monitor in the environment and the frequency of the sampling we require in space and time. Quite evidently, that monitoring must allow us to assess as accurately and precisely as possible the degree of exposure of various targets. It would be logical to see the network of measurements most closely grouped about targets in which we are not prepared to see any significant changes at individual level, while a looser network of monitoring might be acceptable where we are concerned only to prevent changes at the ecosystem level. Moreover, if we are prepared to accept changes at all levels short of the ecosytsem level, biological monitoring becomes evidently acceptable. If we monitor the performance of indicator species, we are tacitly accepting that we are prepared to see their numbers or their performance impaired but because the ecosystem level is unlikely to show change until such impairment at individual level has occurred on a substantial scale, the biological monitoring programme yet gives us something of a safety factor and time span in which to effect corrections should that be needed. On the other hand, biological monitoring may be less acceptable where we are seeking to prevent damage at the individual level unless we can find an organism which is an order of magnitude more susceptible than the targets we are chiefly concerned to protect and displaying a similar sensitivity to the whole interacting matrix of factors. Alternatively, we can monitor at the biochemical level (for example following lead levels in human blood or organochlorine levels in wild life) so as to gain early warning of effects at the organ or individual level. Whatever conclusion we reach, the design of a competent monitoring system must be greatly influenced by our knowledge of the exposure-effect relationship between pollutants and the targets with which we are particularly concerned.

Trends in Pollution and Social Response

Pollution has become a subject of increasing national and international concern in the last two decades. But it is fair to point out that the majority of the incidents which have served to focus this concern have been due to

THE BIOLOGICAL EFFECTS OF CHEMICAL SUBSTANCES 193

localised concentrations of particular pollutants in pollution 'hot spots' (*e.g.*, smoke and sulphur dioxide in the London 'smogs' of the 1950s, mercury and cadmium in water receiving effluent from certain Japanese factories, and raised blood lead levels near a number of factories using lead) It is also fair to add that while the number of incidents causing concern has mounted, the actual damage due to pollution has declined: London 'pea-soupers' and acute pollution near industries almost certainly harmed more people fifty years ago. What has increased is our capacity to diagnose the problems, our will to respond, and our fastidiousness in environmental matters.

There is no evidence of damage to human health or ecological stability on a global scale, due to a globally-distributed pollutant (except perhaps for the statistical prediction of increased radiation damage due to the fallout of the products of nuclear weapon testing in the atmosphere). The most-quoted examples of global trends in pollution, the gradual increase in the proportion of carbon dioxide in the atmosphere, the trace quantities of DDT and other pesticides to be found in oceans and wildlife throughout the world, and the possible changes in atmospheric turbidity due to the injection of fine dust into the air, are all no more than changes in pollutant level without as yet proven consequences for living 'targets'.

Nonetheless, it does not follow that increased emissions of pollutants will not have unwelcome effects, both through the creation of more and more unacceptable local 'hot spots' and through the elevation of the general level of contamination of ocean and atmosphere to the point at which undesirable climatic or ecological changes ensue. Granted that the latter process, in particular, is likely to build up slowly and require an equally long time and great effort to reverse, it is clearly important to improve our scientific knowledge of pollutants and their effects and our predictive capability, as well as our technical capacity to control pollution. In the past, we have allowed pollution damage to develop, and had to cure it when we no longer found it acceptable; in the future we need to forestall such damage and this requires better understanding than we now have.

Trends in Pollutant Emission

It is a historical fact that some pollutants have been emitted in steadily rising quantities, in parallel with mounting human populations and increasing industrialisation. The generation of man's body wastes and food residues, and the body wastes of his livestock, inevitably rise in direct proportion to population. Energy generation through the burning of fossil fuel releases carbon dioxide, carbon monoxide, oxides of nitrogen and of sulphur, water-vapour, hydrocarbons, and a variety of particulates. Total emissions of all these have risen steadily over past decades or even centuries, although man still adds less of many of them to the environment than is contributed by natural processes. There is a similar tendency for the vastly wider range of pollutants generated by industry to rise in direct relation to industrial growth, unless anti-pollution measures are taken, or unless inherently 'cleaner' processes are developed.

But in practical terms many of the most noxious substances—persistent organohalogens, heavy metals, radioactive materials, and known carcinogens —are being released in diminishing amounts and their concentration in the biosphere—in the U.K., if not globally—is falling. This is because of recognition of the need for abatement and the development of increasingly efficient pollution control technology. The pollutants still on a rising trend are generally those not known to be harmful at present or projected levels and hence not yet justifying control. Recognition of the need for control, e.g., on emissions of hydrocarbons, carbon monoxide, and lead from cars, lead from industry, or sulphur oxides from low-level chimneys, has already begun to affect the trends in the products of fossil fuel combustion, until recently looked on as harmless. However, abrupt changes in the cost of various fuels will necessarily have a modifying effect on control policies adopted in future for various forms of energy generation. For we live in a situation where marginal benefits and marginal control costs have to be balanced, and the balance can shift either way.

Factors Influencing Future Trends

In crude terms the concentration of a pollutant in the environment is the result of emission and removal. In practical terms, the important determining factor is man's deliberate control over emissions. This is in turn governed by the costs and benefits of the activities leading to the generation of the pollution and the actual or predicted cost to the community of the damage it does. Research, leading to a deeper understanding of these matters, is likely to be a powerful determinant of the actual future trends in emission, concentration and effects of any substance.

It is impossible to catalogue all major pollutants and predict their future levels. But we can reliably forecast a continuing decline in damaging industrial emissions such as acidic and alkaline vapours, particulates, and heavy metals, and in persistent organohalogens such as PCBs and DDTs, for which substitutes either exist or are being sought. We can predict a slowing down in the upward curve of emission of pollutants from motor vehicles since most developed countries are adopting standards that will halt, or even reverse, this trend within their borders. Substitution of nuclear for fossil energy sources will also lead to a net decline in gaseous and particulate emissions, although imposing a demand for continuing stringent controls on radioactive discharges. Recent international agreements seem certain to lead to a progressive reduction in the dumping or discharge of organohalogens, oils, heavy metals, and known carcinogens to the sea and rivers, and the levels of these substances in the ocean and marine life should soon begin to fall.

Nevertheless, pollutants continue to be an inevitable accompaniment of man's increasingly technological society, and we shall always have wastes to dispose of in the environment. Indeed, it is entirely legitimate to use the environment for this purpose, so long as we do not over-tax its finite capacity to disperse and degrade wastes. With increasingly stringent demands for the testing of new substances for environmental effects before they are used or discharged, and increasingly precise analytical, toxicological research method at our disposal, the probability is that pollutant levels will less and less frequently be allowed to mount to the point at which damage begins.

The Trend in the Cost of Pollution Control

This essentially optimistic projection is grounded on history, for in Britain and most other developed countries, pollution has become a diminishing problem as our awareness, scientific understanding, and technological capabilities have grown. But there is a price to pay for continued optimism. There is a tendency for the energy cost of pollution abatement or prevention to rise disproportionately as populations grow and the standard of living and technological development increase. This is because:

(a) the capacity of the environment to disperse and degrade pollutants is finite, and any increment of pollution over and above that capable of natural degradation demands treatment prior to release;

(b) the larger that increment, the purer the effluent may have to be, volume for volume, if the quality of the environment is to be sustained;

(c) the costs of removing pollution from an effluent tend to mount exponentially as one moves towards greater purity;

(d) the development of substitute or new products becomes more and more costly as the community demands increasingly stringent tests of their safety;

(e) the environmental quality which people demand tends also to rise in parallel with the standard of living, and this provides a 'positive feed-back' increasing the costs listed above;

(f) as we become more and more dependent on technology to purify emissions and protect the environment, so the consequences of break-down become less acceptable and more thorough 'fail-safe' mechanisms are needed, often at substantial cost ($c.f.$ the standards required for nuclear as against conventional power stations).

As economic and technological growth proceeds, an increasing proportion of GNP thus tends to be demanded for environmental protection. If this is to be prevented, increasing effort needs to be devoted to the development of better, safer, and more efficient technologies, giving higher quality controls at less cost. This is one of the needs of society to which chemistry has shown itself responsive, and must continue to respond.

Recycling and Renovation of Water

by M. N. Elliott and D. C. Sammon; A.E.R.E., Harwell

Keeping it Clean

THIS group of three topics is entitled 'Keeping it clean' and the relevance of water in this context is obvious. We use water in our homes for cleaning ourselves and water plays a key role in the maintenance of our current high standards of hygiene. Water is used to take away many of our waste products and yet this same water is eventually returned to our environment. We are now making considerable efforts to keep our environment clean or to make it cleaner than perhaps it has been in the past and the quality of the water returned to the environment is of crucial importance in this respect. It is gratifying to read[1] that many of our rivers have become less polluted in the past few years. The cleanliness of our environment is important not just for the practical purpose of staying alive but also for enjoying life. Our rivers and lakes and reservoirs afford much enjoyment in terms of boating and fishing and just by being there and adding to the view.

In our industrial society the ready availability of water is important both in the manufacturing industries and in the generation of power. With availability must be coupled quality to meet the various needs of industry, and pollution control to keep the environment clean. The simplest approach is to abstract water from surface or underground supplies, pretreat to the standard required, use, treat to control pollution and then return to surface or ground. This same water may, of course, be used again—perhaps further downstream—and in this way water recycle and renovation are practised. However, this paper will be concerned with recycle and renovation on a smaller, essentially local, scale.

In some instances additional supplies of water are not readily available and attention must be directed towards more effective use of existing supplies. In others, economic factors favour re-use of water. It may be cheaper to renovate and re-use than to pay for water and for the costs of treatment prior to dis-

197

charge. If extensive treatment is required to limit pollution it may cost little more to produce water of quality suitable for re-use. Finally, recycle of water on a local scale—within the factory or even the process line—may permit the recovery of valuable materials.

These pressures, then, are leading to increased use of recycle and renovation and in the future this trend is likely to be maintained.

Availability and Use of Water

The amount of water falling as rain (and snow) on this country is considerably in excess of our requirements even allowing for the fact that, of the total falling, only about half reaches the sea *via* lakes and rivers. The other half is lost by evaporation. The quantities abstracted for various purposes are shown in Table 1.[2] The figures refer to 1971 and are likely to have increased by about 3% per year since then.

Table 1: Quantities of Water Abstracted in England and Wales during 1971 (From Water Resources Board 9th Annual Report)

	Quantity abstracted per day (units of 10^6 m^3/day)
Public Water Supply	14.3
C.E.G.B.	18.9
Industry	9.2
Total	42.4

Water abstracted by the CEGB is mainly for cooling. Industry uses water abstracted directly and a somewhat smaller quantity drawn from the public water supply (about $\frac{2}{3}$ *i.e.* 6×10^6 m^3/day). Agriculture, including irrigation, accounts for only a small daily volume, namely 0.24 of the total of 42.4. Almost all of the water used for domestic and much of that used for industrial purposes is discharged to a sewer and treated before being returned to surface or ground. The rest is discharged with appropriate treatment to sea, estuary, or river, or is lost by evaporation. Some of the water required for cooling in electricity generation is used on a once-through basis, but it is much more common to use a recirculation system in conjunction with cooling towers. For example, in a 2,000 MW power station, the recirculation rate would be about 5.5×10^6 m^3 per day, but the make-up would be about 1% of this figure and would be needed to replace losses by leakage and evaporation and to control the concentration of dissolved salts.

Though the total demand is a small fraction of the total volume available this is not necessarily true in any particular locality. Water supply problems are more acute in the heavily populated and industrialised areas of the country than elsewhere. Likewise water availability shows a seasonal variation. Various schemes have been proposed to even out these variations in water availability.

The above comments, of course, only refer to quantity, whereas quality is obviously of extreme importance. The seas and oceans offer a virtually inexhaustible supply of water, but of a quality which is totally useless for most applications.

Water for Different Applications

The various industries have differing water quality requirements depending on the use made of the water. Typical uses are summarised in Table 2; for these various applications water quality varies considerably from simple cooling or quenching where raw water from almost any source is adequate

Table 2: Industrial Uses of Water

Steam generation	Power or process steam
Cooling water	Steam condensation Other cooling needs including quenching
Process water	Washing medium Transport fluid Solvent or suspending medium Reactant chemical Essential component of food or drink

to the manufacture of semiconductors where the levels of dissolved and suspended solids are so low as to merit the term ultra-pure water.

The water quality required for steam generation varies with the pressure of the boiler used. For steam pressures above 1500 lb/in² demineralisation is needed down to total dissolved solids levels of less than 1 mg/l. Wherever practicable most of the condensed steam is returned to the boiler to minimise the make-up volume. This is, of course, not possible in general for process steam.

Cooling water can be used once-through or in a recirculation system and the principal quality requirements are dictated by the need to limit corrosion and scale formation. For quenching of metals or slags only low quality water is needed and can be re-used many times with intervening settling ponds.

For the process uses, requirements vary from potable quality for water incorporated in food or drink to the ultra-pure needed for semiconductors. In rinsing operations only the final rinse may have to be of high quality. However, for those applications generalisations are of little use; each has to be considered in the light of the specific requirements.

Treatment of Waste Water

The availability and cost of water of adequate purity represent one constraint on the user and a second equally applicable one arises from the problems of disposing of the water after use. In principle, any particular standard for feed or effluent can be achieved at a price, thus specifications and costs are heavily interdependent. It is useful to consider standards for effluents before going on to discuss the economic aspects of recycle.

Table 3: Typical Consent Conditions for the Discharge of Industrial Effluents to Municipal Sewers

Substance	Maximum allowed
pH range	6—10
Caustic alkalinity (as $CaCO_3$)	2500 mg/l
Sulphate (as SO_3)	1000 mg/l
Free ammonia (as NH_3)	500 mg/l
Suspended solids	1000 mg/l
Tarry and fatty matter	500 mg/l
Sulphide (as S)	10 mg/l
Immiscible organic solvents ⎱ Petroleum and petroleum spirit ⎰	nil
Calcium carbide	nil
Temperature	45°C
Chromium (as Cr)	50 mg/l
Copper (as Cu)	50 mg/l
Separable oil and/or grease	400 mg/l

At an inland site the choice lies between discharge to surface water or to public sewer, assuming there is adequate treatment capacity at the sewage works. Authorisations are required in both cases and consent conditions are specified. Typical consent conditions for discharge of industrial effluent to municipal sewers are listed in Table 3.[3] These are designed first of all to protect the sewers and sewermen by preventing corrosion, explosion,

RECYCLING AND RENOVATION OF WATER

Table 4: Main Classes of Pollutants

Toxic or harmful substances	Pesticides and herbicides from agriculture and horticulture
	Radionuclides from laboratories and hospitals
	Acids and alkalis, for example pickling wastes
	Heavy metals—Cu, Zn, Ni, Cr, Cd, Pb which can be toxic at <1 mg/l
Substances which increase the oxygen depletion of the water	Chemical reducing agents
	Biologically labile substances, for example sewage
	Surface active agents which interfere with the solution of oxygen from the air
	Warm water, which reduces the saturation concentration of oxygen and also increases the rate of certain processes
Substances which are indirectly harmful	Chemically inert suspended solids
	Colouring agents
	High concentrations of soluble inorganic salts

blockage, or build-up of toxic fumes. The second purpose is to minimise interference with the sewage treatment process. Some toxic materials inhibit biological treatment while other substances are not affected by the processes normally used.

The restrictions imposed on the discharge of industrial effluents to surface waters are based on the effects to plant and animal life and to the amenity value. The main classes of pollutants are listed in Table 4 and need little further explanation. Typical consent conditions are given in Table 5,[3] for fishing streams; for non-fishing streams some of these are relaxed. These consent conditions use the basic 30:20 standard (30 mg/l suspended solids,

Table 5: Consent Conditions for Discharge to Streams

Constituent	Maximum Allowed
BOD_5	20 mg/l
Suspended solids	30 mg/l
Sulphide (as S)	1 mg/l
Cyanide (as CN)	0.1 mg/l
As, Cd, Cr, Cu, Pb, Ni, Zn, individually or in total	1 mg/l
Free chlorine	0.5 mg/l
Oils and greases	10 mg/l
pH	5—9
Temperature	30°C

20 mg/l Biological Oxygen Demand) recommended by a Royal Commission in 1912 and still widely applied today as long as the waste is diluted more than eight-fold in the river.

Cost of Water

Abtraction or supply costs for water vary considerably over the country. Until recently much of the water was supplied from reservoirs built as much as 100 years ago and the cost of water from these contains little or no allowance for depreciation. In recent years expensive reservoirs have been built and water from them will cost more. In 1970 the average price of water was 3.3p/m^3 (15p/1000 gallons)[4] but in new development areas the cost may be as high as three times this amount. To this supply or abstraction cost must be added the cost of any treatment required to meet the consent conditions. For discharges to municipal sewers, costs are calculated according to a formula and depend on the characteristics of the waste and on local circumstances. The average cost of sewage treatment[5] is about 1½p/m^3, but the charges for treating industrial effluent can be considerably less. One cannot generalise about the costs of treatment before direct discharge to a river, but there is no reason why these cannot be lower than charges for for municipal treatment. Municipal plants are built to last a long time; industrial treatment plants can be tailored for the particular conditions and designed to last as long as the waste is likely to be produced.

The cost of water to the user is thus the sum of supply and treatment costs and can be as high as 6.6p/m^3. This total cost is an essential part of processing and manufacturing costs and should be kept to a practical minimum. It is clearly good business sense to use water more than once and to treat and re-use waste water. However, to make the necessary decisions it is essential to be aware of the water demand. Considerable economies can be achieved just by drawing up a water balance sheet and keeping all requirements to a carefully controlled minimum. Instances have been quoted[6] where a considerable fraction of the water supplied could not be accounted for. Elimination of such wastage is fundamental to water economy. Various other practices are being used, such as using the effluent from a final rinse bath to feed a previous rinse and so on, instead of using fresh water for each bath. However, these and other methods, valuable though they are, are outside the scope of this paper. In many cases considerable savings can be achieved by recycling water.

A simple cost calculation is quoted[7] where a once-through cooling system is replaced by a recirculation system with cooling tower. Here the capital cost of the tower and associated equipment was offset by the savings in water costs and the time required to recover the additional capital expenditure was just under five months. Water re-use is already extensively practised, but the extent varies widely even in one sector of industry. For example, the amount of recycle has been found[8] to vary between $2\frac{1}{2}\%$ and $82\frac{1}{2}\%$ for individual copper fabricators in the U.K. These variations stem from local variations in water costs and also sometimes simply from a lack of appreciation of the benefits that can be obtained.

Recycle and Renovation

Use of water is usually accompanied by deterioration in quality; thus most recycle schemes require some sort of renovation process to bring the quality back to that required. The cost of the renovation has to be met from the savings resulting from decreased water usage. Sometimes the decision between once-through and recycle is simply based on how much additional processing is needed to convert water suitable for discharge into water of quality suitable for re-use. In a recent article[9] the criterion applied was that this additional cost was more than offset by the savings in water costs. Two new installations were described; one a plant for producing Instant Mashed Potato and the other a bottle washing plant. The wastes in both cases were high in Biological Oxygen Demand and had to be treated to the Royal Commission standards mentioned earlier. Primary and secondary treatment stages were required to meet the standard and the tertiary stage was needed to give water suitable for re-use. The cost of this tertiary stage was justified by the lower demand for water.

Recycle and renovation are not without problems. Recycle may mean that heat is not carried away and this may cause secondary problems as reported[10] for a paper-mill where the higher water temperature caused condensation and the growth of slimes which though they did not affect the basic process, caused operational problems. If water is used in succession in various steps in a recycle loop these various steps become linked to each other and failure or troubles in one step may cause problems elsewhere and even lead to the whole cycle having to be shut down.

Renovation Processes

A wide range of renovation processes is available and one or more would have to be used to meet the particular requirements. These vary from the simplest ones such as filtration and screening to more sophisticated (and expensive) such as the desalination techniques studied and developed in the

Table 6:

Constituent	Treatment Process	% Removed	Cost
Suspended solids	Screening	90	A
	Sedimentation	60	A
	Microstraining	60	A
	Filtration	70	A
	Flotation	60	B
	Coagulation	80	B
Oil	Settling	95	A
	Filtration	90	A
	Absorption	30—80	B
	Flotation	90	B
Oxidisable soluble organics	Stabilisation basins	50	A
	Percolating filter	60	B
	Activated sludge	60	B
	Aerated lagoon	50	B
	Anaerobic oxidation	50	C
	Activated carbon	70	C
Soluble inorganics	Ion-exchange	90	D
	Electrodialysis	90	D
	Reverse osmosis	90	D
	Freezing	80	D
	Distillation	95	D
	Liquid–liquid extraction	80	D

Costs: A <0.5 p/m³   or <2.5 p/1000 gal.
      B 0.5—2 p/m³   or   2.5—10 p/1000 gal.
      C 2—4 p/m³   or   10—20 p/1000 gal.
      D >4 p/m³   or >20 p/1000 gal.

U.K.A.E.A. In the space available a total review is impossible, but various processes will be mentioned for removing various pollutants and costs will be given in general terms. More exact costs depend critically on the scale and nature of the particular application, thus the figures given should only be taken as a general indication. One product of all these renovation processes is, of course, water suitable for recycle, but there are almost always other products such as sludges and concentrates. The problems associated with disposal of these unwanted products must also be solved and the costs must be included in the overall balance.

Approximate treatment efficiencies and costs for renovation processes have been tabulated by Sawyer[11] and much of this information is used in the following discussion. The first class of processes to be considered covers removal of suspended solids and immiscible liquids and the details are given in Table 6. Coagulation and flocculation are effective in removing finer solids than is true for the other processes except flotation. A wide variety of coagulants is available and the process can be used for the removal of heavy metals and also for the removal of oxygen demand as in the newer physico-chemical treatments for sewage. For the removal of colloidal species ultra-filtration can be used.

A large number of soluble organic species can be destroyed by the bio-logical oxidation processes listed in Table 6. Percolating filters include the older type using beds of rock or coke as well as the newer ones using plastic media. The latter take up less area. This type of process is used for sewage treatment and also for the much more polluting wastes from food processing, paper and pulp processing and the manufacture of textiles. Frequently several stages are needed to produce adequate removals. Adsorption on activated carbon is used as a polishing step but can also be effective in removing colour and species that are not readily oxidised.

Biological processes and ion-exchange in the $0.5—2p/m^3$ range can be used for the removal of ammonia, nitrogen, or phosphorus.

For the removal of soluble inorganic species the whole range of desalina-tion techniques shown in Table 6 is applicable. These processes are the most expensive but can give appreciable removals of other species as well. Many of the processes listed earlier have little effect on, or increase, the salt content and extensive use of recycle can cause a build-up of salt which can be controlled using a desalination process. With repeated use the salinity of rivers is increased but so far this is not troublesome except in isolated instances. Ion-exchange and electrodialysis result in the removal of ionic species whereas reverse osmosis, freezing, and distillation are effective for a wide range of solutes. Liquid–liquid extraction is likely to be most useful for the removal of specific solutes.

Many of these processes are well established but some like reverse osmosis are just beginning to find wide application. There is thus no shortage of renovation technology; the controlling factor is the cost.

Recycle of Other Materials

As mentioned earlier, water is the principal product of renovation, but there are other products and some of these are valuable. As increasing emphasis is placed on conservation of resources, the recovery of materials must assume greater significance. One example of this is the treatment of waste rinses from chromium plating.[12] The chromium could be removed by precipitation, but an alternative is to use ion-exchange and reverse osmosis to remove this species from the rinse and return it directly to the plating bath. The treated rinse water is further processed and re-used in various metal-plating operations while the cost of the chromium recycle loop can be recovered in about one year due to the value of the recycled chromium.

Future Trends

Many references have been made to sewage treatment in this paper, but so far there seems little likelihood in this country of recycling treated sewage into the public water supply, except *via* rivers and after dilution. Though recycle would probably be uneconomic at the moment public opinion may well be more important than economics. Moves in the direction of recycling of sewage are being made in Holland[13] and South Africa.[14]

The types of recycle described earlier are likely to be used more and more. The demand of water will increase as the population increases, as water use per person increases with the standard of living and as industrial use increases with increased production. Water costs will rise as more expensive sources have to be used and it seems likely that more stringent pollution control will result in more stages of treatment. With further development of renovation technology costs of individual processes may decrease slightly. All these factors point to increased use of recycle and renovation in the future.

Bibliography

[1] River Pollution Survey of England and Wales—updated 1972. Department of Environment and Welsh Office, HMSO, 1973.
[2] 9th Annual Report of Water Resources Board, HMSO, 1972.
[3] P. G. C. Isaac: *The Chem. Engineer*, 1973, No. 273, p. 253.
[4] Pay and Productivity in the Water Supply Industry, Report No. 152, 1970, National Board for Prices and Incomes.
[5] 'Water Statistics 1968—69', The Institute of Municipal Treasurers and Accountants.
[6] R. K. Chalmers: *The Chem. Engineer*, 1974, No. 282, p. 94.
[7] G. B. Hill: 'Industrial Pollution Control', ed. K. Tearle, Business Books Ltd., London, 1973.
[8] E. C. Mantle, R. Savage: 'Water Economy in the Non-Ferrous Metals Industry', BNFMRA, London, 1970.
[9] T. S. Allen: *Trans. ASME*, November 1972, p. 1094.
[10] D C. Morris: *J. Water Pollution Control Fed.*, 1973, 45, 1939.
[11] G. A. Sawyer: *Chem. Engineering*, 1972, 79, No. 16, p. 120.
[12] J. Schrantz: *Ind. Finishing*, 1973, 49, No. 6, p. 38.
[13] Private communication.
[14] O. O. Hart, G. J. Stander: 'Applications of New Concepts of Physical-Chemical Waste Water Treatment', Conference at Nashville, Tennessee, September 1972.

The Role of the Chemist in Keeping the Air Clean

by J. S. S. Reay, Warren Spring Laboratory, Stevenage, Herts.

Introduction

IT is all too easy to identify as air polluters the chemist together with his brother the engineer who has translated his reactions to the scale of tons, thousands of tons, and even millions of tons, and thus given us most of the materials on which our modern way of life depends. In the van of the industrial chemical revolution of the nineteenth century, the UK was not only early in the field in creating air pollution from its many industries but also first to introduce a regulating inspectorate to deal with the worst excesses of its chemical activities. One has only to read the Alkali Act's list of scheduled industries with its occasional archaisms like muriatic acid to appreciate how long-standing is this system of control and how intimately it is bound up with the chemical industry.

If one looks further at the list of scheduled processes one finds that it includes many which members of the general public do not immediately identify with the chemical industry, although it is frequently the chemical transformations which they involve that are the potential sources of environmental pollution.

Installations producing pharmaceuticals, pesticides, detergents, or fertilisers are clearly regarded as part of the chemical industry. Monomers, polymers, and plastics are recognised as the products of chemical reactions upon oil, but paper, metals, and cement are not immediately identified by the public with chemistry. Even less does the man in the street associate an electricity works with chemistry nor, for that matter, his own domestic combustion and all the fuelled vehicles on our roads and railways.

There is great diversity in the large volume of our activities which are potential polluters and are at the same time vital to our prosperity and our way of life. The main contributors to our national air pollution picture are combustion and the fuel and power industries and I want first to devote some

207

time to these and to show where the chemist can contribute in reducing pollution.

It is wrong to date our concern with combustion pollution back only to the great London smog of 1952 although it is true that much of our clean-air legislation was initiated by that disaster. We had already abandoned our romantic attitude to the London pea-souper but we had not generally thought of it as a mass killer, and only over the intervening years have we pieced together the story of chronic health effects associated with sulphur dioxide and smoke pollution. Even before the Second World War the chemists in the UK had turned their attention to scrubbing sulphur dioxide from the flue gases of urban power stations. Wet scrubbing techniques are not without substantial drawbacks such as the behaviour of a cooled, wet plume and the problem of secondary effluent or waste product disposal. Today, nearly forty years on, flue gas scrubbing is applied only in special circumstances.

By and large we have chosen to adopt a purely engineering solution to reducing the contribution of major combustion plants to ground level sulphur dioxide concentrations. We have made general use of chimneys whose heights are decided on the basis of the expected sulphur dioxide output and the extent to which that sulphur dioxide will be dispersed and diluted before reaching ground level. This high-stack policy, adopted too by many other countries, has met its declared objectives well.

However, in certain situations, there is an increasing call for the reduction of sulphur dioxide emissions. In Japan, for instance, the view is held that the coastal strip of intense industrial development has such large sulphur dioxide emissions that the distributing power of high stacks is just not enough. Then, too, there is concern over the long-range drift of sulphur dioxide pollution and the possible harmful effects of it being brought to earth in regions which cannot accept the acidity without damage. A study of the magnitude and mechanisms of such pollution transport, over distances of up to 1,000 km and more, is being undertaken by a group of NW European countries within OECD. A number of studies are also being made of the chemistry of the sulphur compounds in the air. Some countries are already legislating for reductions in sulphur dioxide emissions. The obvious expedient of using cleaner fuels is limited in various ways and it is to the chemist that they must turn for sulphur removal techniques, particularly those which are cheap and reliable.

One possible line of attack is fuel desulphurisation. Crude oils from various fields have a variety of sulphur contents and these are reflected in the sulphur contents of some of the products made in the refineries. Some of the lighter products are desulphurised pretty universally to quite a low sulphur content, but the heavier residual fuel oils have not normally been specially desulphurised, and it is these products which have been used in many medium and large combustion installations. The chemist has already developed some desulphurisation techniques for these heavier fractions, but the extent of their application should logically depend on their cost relative to the benefits to the environment which may be shown to be achieved by their use, rather than simply high stacks with their attendant costs.

The results of air pollution monitoring in the UK, considered in conjunction with fuel statistics, have shown that most of the ground-level sulphur dioxide concentrations in our towns are produced by combustion plant with low chimneys, and if we seek improvements in our urban pollution levels it is the fuel for these installations that we should look at first.

In the UK we still generate most of our electricity from coal. Here fuel desulphurisation is much more difficult. If the coal is suitably pulverised a good proportion of the pyritic sulphur can be removed by physical separation but the organic sulphur yields only to more sophisticated techniques such as solvent extraction. So more research emphasis is rightly given to methods of removing sulphur dioxide either during combustion or from the subsequent flue gases. Considerable work has been done by the National Coal Board and others on fluid-bed combustion/desulphurisation of coal and oil. The strict legislation being progressively implemented in the USA is promoting greater efforts on flue gas desulphurisation, especially by cyclic systems not employing vast amounts of water. There is plenty of scope for the chemist here and especially, in my opinion, if he can come up with an economic scheme for moderate sizes of combustion plant. Methods of desulphurisation must beware of merely transferring the pollution problem from the air to some other medium, and the chemist should look also into the problem of waste utilisation—perhaps a sulphur-based polymer—since large-scale desulphurisation would produce vast amounts of by-products.

Another area of sulphur pollution calling for the chemist and engineer is combustion control to avoid the production of sulphur trioxide.

The Clean Air Act of 1956 was directed more than anything else to the control of smoke emissions. The last twenty years have seen big improvements in the arrestment of non-combustible particulates and in straight combustion efficiency with the emission of less carbon monoxide, partly burned fuel, and smoke. In the traditional domestic grate bituminous coal burns with a smoky flame; chemists have played an important role in the production of solid smokeless fuel from coal. More recently combustion engineers have developed devices for the smokeless burning in the home of certain grades of normal bituminous coal.

Attention is turning increasingly to the control of emissions of oxides of nitrogen from combustion because of the secondary pollutants which they can form by interaction with other materials in the air. Considerable research is being done on the kinetics of flame reactions. While some of the nitrogen may come from the fuel itself, the main source is the air, and modern high-temperature flames cause substantial formation of NO.

Motor Vehicles

So far I have been dealing primarily with stationary combustion plant. Motor vehicles—both spark ignition (SI), and compression ignition (CI)—are also the object of much antipollution research. Again the chemist is well to the fore. Vehicle engines differ markedly from normal furnace equipment in the large surface-to-volume ratio and the cool condition of the combustion chambers. The SI and CI engines differ from one another in the propagation and type of flame and in the fuel-to-air ratios. While many of the improvements which can be made are achieved by engineering modifications, these modifications are based on a knowledge of combustion chemistry.

A number of countries have introduced improved standards for smoke emissions from diesel engines and carbon monoxide and hydrocarbons from petrol engines. The call for greater reduction in emissions of carbon monoxide, hydrocarbons, and oxides of nitrogen has initiated an enormous amount of chemical research on flame propagation and kinetics in SI and CI engines and on catalytic systems and thermal reactors for exhaust gas treatment. The chemical situation is extremely complex with the relative kinetics of competing reactions an important feature.

Nearly forty years ago the combustion chemist made a major contribution to fuel efficiency with the commercial introduction of anti-knock additives. Since that time there has been an enormous increase in the use of these additives and lead residues have become widely distributed throughout the environment. The concentration of airborne lead in busy streets and the lead content of city dusts has led to concern in some quarters about their possible effects on health. A number of countries have introduced timetables for the reduction of lead content in petrol. Those countries intending to use catalytic devices for the reduction of gaseous emissions are also keen to remove lead from petrol because it poisons the otherwise most promising catalysts.

The maintenance of octane rating without lead calls for a different hydro-carbon composition for petrol. This implies more fuel-consuming re-forming at the refineries and the greater use of aromatics which, it is argued, might produce potentially harmful hydrocarbon emissions in vehicle exhausts.

There are undoubtedly places in the world where the density of vehicle emissions is so high as to produce significant adverse effects. The most striking example of this is probably the Los Angeles smog phenomenon where the aerial interaction of oxides of nitrogen and unsaturated hydrocarbons under the influence of sunlight produces ozone. Ozone is itself damaging to certain vegetation when in sufficient concentrations but further reactions of ozone, free radicals, and oxides of nitrogen go on to produce other odorous and irritant compounds. The chemist has done society a considerable service by elucidating these complex reactions and showing which emissions must be reduced when it is desired to improve the situation.

Most of the schemes so far proposed for reducing vehicle pollution produce disbenefits in the form of increased fuel consumption. There is great scope for the development of low-polluting systems with high fuel efficiency. This may involve quite radical departure from the traditions of internal combustion engines.

If, as we must assume, our future energy requirements for heat and electricity will increasingly be met by nuclear energy rather than by fossil fuels, we shall still have the problem of fuel for transport where electricity cannot be the complete answer, certainly not for aircraft. Thus there is already some effort on systems for producing and utilising hydrogen and synthetic fuels such as methanol, ammonia, hydrazine *etc.* Considerable chemical ingenuity is being shown in seeking alternative ways of obtaining hydrogen at low cost from hydrocarbons or even water.

Nuclear Power

Power produced by nuclear energy offers to do away with the conventional combustion pollutants (except heat) but there is concern not only over reactor safety but over the hazards of nuclear fuel processing and the disposal of radioactive waste. Thus the chemists' advances on these latter problems can, at second hand, improve air pollution by hastening the safe transition to wide use of nuclear power generation. The sort of thing I am thinking of is the introduction of frits containing radioactive elements in so unreactive a structure that the release of these wastes from sealed containers would not necessarily imply their serious dissemination.

Fuel Availability and Costs

No apology is required for the emphasis in this paper on combustion products. It reflects appropriately the importance of this form of pollution. This paper was drafted before the recent dramatic rises in world oil prices. It is interesting to give preliminary consideration to the effects these changes may have on environmental protection activities. There are two important fuel parameters —one is price and the other availability.

In the short term, shortage of fuel, at least of preferred quality, must lead to the use of more polluting grades. Although it may not always prove practicable it will be desirable from the environmental point of view to concentrate the use of dirtier fuels in devices with high stacks. Where dirty fuels have to be used in low chimney situations it would be attractive if this could be done preferentially in meteorological conditions not conducive to the build-up of high ground-level concentrations of pollutants. This would, however, be difficult to organise.

In this time-scale it is not likely that the chemist can make any contribution to pollution abatement but, if shortages continue and are combined, as seems inevitable, with much higher fuel costs, there will be considerable economic incentives to reduce fuel usage which will in turn, reduce pollution. Improved fuel efficiency is the first obvious goal; then there is better heat transfer and greater use of waste heat, e.g. in district heating. Improved insulation in houses and other heated buildings will be attractive as fuel costs soar. At second hand, the chemist providing foam polymer for insulation is thus capable of reducing air pollution.

For the last ninety years we have, by and large, been elaborating a few basic systems of energy generation from fossil fuels. Great strides have been made in engineering realisation, e.g. in gas turbines, but mainly within the framework of cheaply available hydrocarbon fuels. Only in the defence and space fields has research been unconstrained by normal economic considerations. It is ideas of this breadth and originality which we shall need increasingly so that, if not in this decade, at latest by the end of the century we can have the necessary range of power systems to match the diminishing reserves of fossil fuels. These reserves will, in any case, be more than ever valuable as the basis of the 'synthetic' industries.

The Oil Industry

Turning from combustion to the fuel industry, I have already referred to fuel desulphurisation which, for oil, would presumably be carried out at refineries.

Current refinery processes are not without their polluting effluents, particularly sulphur dioxide and odours. A point to be remembered is that sulphur contained in crude oil has got to end up somewhere. That which is not exported in fuel products will in part be burned within the refinery and in part be recovered as sulphur in the desulphurisation of lighter oil fractions. A further quantity, infinitesimal in comparison, is contained in a range of sulphur compounds some of which are particularly smelly. 'Best practicable means' for refineries burning sulphurous fuels is the use of a stack high enough to produce adequate dispersion. Some sulphur compounds are burned in a flare stack. Sulphur from desulphurisation ends up, by and large, as by-product elemental sulphur. If large-scale desulphurisation of heavy fuel oils were undertaken we might expect not only huge further tonnages of by-product sulphur but also further use of fuel within the refinery to run these processes and so further emissions of sulphur dioxide and other combustion pollutants. General fuel economy throughout the country offers at least a reduction in the rate of growth of refinery operations but there may be side effects if there is not a corresponding trend in the requirements for other refinery products, and thus changes have to be made in the extent of cracking, re-forming *etc.* The chemist again has a big field for the application of his ingenuity in refinery processes.

It would be impossible here to cover most of the industries with polluting effluents. I shall mention only a few major ones which exemplify the situation.

Brick Manufacture

As far as SO_2 is concerned, the off-gases from fletton brick manufacture come into the same category as those from power generation in that dispersion from a tall chimney is regarded as representing 'best practicable means'. But these off-gases also contain small amounts of fluorine compounds and even smaller quantities of pungent compounds which produce the characteristic odour downwind of the brickfields. Under fumigation conditions—the odd occasions when the chimney plume comes to earth with minimal dispersion—the gases are capable of damaging vegetation over small areas. Despite intensive study this is an industry where chemical technology has yet to provide an alternative to dispersion which would not be economically impracticable.

Metal Production

The metallurgical chemists' skills in relation to production efficiency can produce problems on the side. Thus the enormous increase in the use of oxygen in various steel-making and refining processes has brought with it the problem of containment of oxide fume—particles ranging from 0.05 to 1.0 μm—perhaps the industry's most difficult gas-cleaning problem. It would be wrong to suggest that the metallurgists made a retrograde move. The advantages of oxygen steel-making outweigh the difficulties of fume arrestment but it has to be said that the steel industry here and overseas did not develop adequate fume abatement until some time after introducing the new processes. It is this sort of piecemeal approach to process design which is becoming less and less acceptable.

One of the major polluting processes, on a world scale, is the smelting of sulphide ores. In the UK we have moved away from such smelting for a variety of reasons and, to a large extent, we import crude metal. Where we still smelt sulphide ores, as in the cases of zinc and lead production, the Alkali Inspectorate interpret 'best practicable means' as implying the removal of sulphur dioxide from the off-gases. This is usually converted into sulphuric acid. At some major smelting centres, such as that in Ontario, Canada, the scale of operations is enormous. To date that industry has been permitted to discharge the bulk of its sulphur dioxide—over two million tonnes a year—straight to the atmosphere through a tall chimney. This is equivalent to ten to fifteen major UK power stations being sited close together. To turn all this ore-derived sulphur dioxide into sulphuric acid implies an annual output of well over three million tonnes of acid, and the general adoption of this solution would certainly produce an embarrassing amount of sulphuric acid. As a result there is considerable interest in the possibility of extractive metallurgists coming up with less polluting alternatives to the traditional pyrometallurgical processes. Hydrometallurgy is thought possibly to offer less polluting possibilities. Personally, I have doubts about this being a general panacea. If the sulphur is there in the ore then it must be contained in some by-product. It is thus not immediately apparent that the new by-products of hydrometallurgical processes will be much easier to dispose of in vast tonnages than sulphuric acid or elemental sulphur. Again, one hope is a useful product from sulphur. In terms of overall energy consumption the pyrometallurgical approach may still be the preferable route to extraction for many ores.

Lead and zinc production not only involves sulphur dioxide but also fine particulate metal generation. Very high efficiencies of abatement are required and achieved on the gases carrying these small metal particles because some, such as lead and cadmium, are recognised as extremely hazardous. Nonetheless, for the largest sizes of smelter/refinery complexes now being built, the total emissions of such metals are substantial. Despite the obvious attractions of economies arising from the scale of such installations, there may have to be limits to the size of plants operating certain types of polluting processes which can be sited in one place because there are technical rather than economic limits to the abatement of their emissions.

Pollution Control Regulations

A lot has been written and said about the British system of control of polluting emissions from industry. Scorn has been poured on the words 'best practicable means' and upon the absence of statutory limits for emissions of most pollutants. On the other hand the Alkali Inspector is never really ultimately satisfied with any level of polluting emission. As soon as a *better* practicable means appears on the scene he will adopt it as his requirement for new installations. Improvements and enforcement are best achieved with the co-operation of industry. Legal confrontation is in itself no solution to a technical problem. I know that the Chief Inspector values highly his flexible position and would not willingly give it up for the apparent simplicity of the legalistic and rigid schemes which some other countries are seeking to impose on their industries. It has been argued that the British system is too passive and that there is no incentive for the industrialist in this country to seek improved abatement technology, particularly if it might prove more expensive. But the record shows that the Alkali Inspectorate has not been purely passive. In many industries the Inspectorate has pressured the operators to experiment with, and to develop, improved abatement systems. Obviously the pressures thus bearing on our industry are not as strong as those now being applied in some other countries and those pressures abroad will undoubtedly produce improvements in some areas—improvements which may then be adopted as 'best practicable means' in this country. What we may miss out on is the actual innovation and the business it may produce; but is not that an incentive to UK industry to be active in this field anyway?

Overall Approach to Process Emission Control

In the short term, existing plants have abatement equipment tacked on to improve emissions, but in the longer term the best way lies in flowsheets designed from the start to produce low emissions. This is a continuing process. Not many years ago a contact sulphuric acid plant was expected to have 98% conversion of SO_2 into SO_3 and the effluent chimney was designed for the corresponding effluent.

The double contact process now provides 99.5 to 99.7% conversion at an increase in capital cost of 10—15%. It has been adopted by the Inspectorate as representing 'best practicable means'. In nitric acid manufacture the chemical engineer is faced with the Inspectorate's requirements for a maximum of 2700 ppm NO_x in the exit gases through a stack such that the 3-minute mean ground level concentration is about 0.16 ppm in average wind conditions. Recognising that the practical limits relate to absorption efficiency the designers have been looking at the relative attractions of more absorption capacity, higher absorption pressures and even the use of oxygen rather than air for the oxidation of ammonia. In respect of chlorine and alkali manufacture one tends to think first of the mercury content of liquid effluent but there are also emissions to air of chlorine and of hydrogen contaminated by mercury. While chemical ingenuity has been applied on the one hand to the trapping of these noxious materials from mercury cells, attention has also been given again to the diaphragm cell as an alternative which no longer involves mercury and thus completely obviates mercury-containing effluents.

Odours

With the cleaning-up of materials which are directly damaging to health, society turns its attention increasingly to other features of industry which reduce environmental amenity rather than create a direct hazard in the physical sense. Noise and offensive smells are in this category. In fact both of these types of pollution can be injurious to health through the stresses produced by their unpleasantness.

Reference has been made to several of the scheduled processes, e.g. brickworks and refineries, where odour emissions are a problem. However, Local Authorities are responsible for control of the bulk of odour-producing processes, especially farming, food processing, and animal by-products; the long recognised 'offensive trades'. In 1971 the Department of the Environment set up a Working Party to examine these problems and to make recommendations about the 'best practicable means' for their minimisation and suppression. The first tasks of the Working Party have been to analyse the problem, to describe how the odours arise, and to survey the known methods of measurement, analysis, abatement, and counteraction. In support of the Working Party, practical work has been set in hand at Warren Spring Laboratory on measurement and abatement techniques.

I have seen it written that the main problem lies in the absence of satisfactory means of measuring odours and the fact that their assessment remains subjective. This does not seem to me to be the main problem at all, although the seriousness of an odour problem undoubtedly depends on the response of individuals, which can vary widely. To my mind the main problems lie in the nature of many of the industries which produce odours and in the fact that the perception of odours is roughly related to the logarithm of the concentration of odorous material. Because the odorous compounds are usually present only in small concentrations in the original gas stream, the technical problems of abatement are much more severe than for most normal emissions; for instance, to reduce the perceived intensity of an odour to one-eighth implies the removal of something like 99.99% of the compounds responsible for the smell.

There are thus real challenges to the chemist. A number of alternative approaches exist. Most smells can be destroyed by burning but the cost of specially raising large volumes of gases to the necessary elevated temperatures can be high. The ingenuity of the catalyst chemist can, in some instances, greatly reduce the required temperatures. For certain smells adsorbents can be found and the application of ultraviolet light or ozone may sometimes be worthwhile. In the use of absorption or scrubbing, the chemist and engineer can contribute in terms of reagents and in obtaining high efficiency of gas/liquid contact.

The solutions to odour problems do not all lie through these removal or destruction techniques. There is no doubt that there is considerable room for systematic improvements in the processes themselves and in the reduction in the escape of odours at various points. These odour problems are taken very seriously by the neighbouring residents exposed to them and it is extremely important that solutions be found, otherwise there is an understandable but far-reaching tendency to refuse permission for these industries in a growing number of localities. This is not a true solution because these processes have to be undertaken somewhere, and there are cogent reasons why they should be carried out adjacent to the point of production of their raw materials.

Conclusions

I have devoted almost all this paper to the polluting processes themselves and to the role of the chemist in controlling and abating polluting emission. The chemist has another important part to play in telling us how clean we need to keep our air. In addition to epidemiological studies involving mainly medical experts, biochemical studies on various natural systems are necessary to elucidate the mechanisms by which pollutants can affect life. For some pollutants there may be threshold values below which they do not damage various systems. For others almost any concentration will have a harmful effect on someone or something. While lack of information on dose–effect relationships and damage costs *etc.* is no excuse for complete inactivity in reducing pollution, rapid provision of effects data is important to resolve the many debates on the basis of fact and to provide rational targets for legislators and process designers.

It is clear that the economic equations of the past did not take into account as much as we now wish, the effects of industrial operations on the environment, nor were the scale of these operations and their overall effect so great and so widespread. We have rightly become concerned about the ecological or general interactive effects of our activities. Before, however, we become too obsessed with all the effects of human activity on other natural systems, let us remember, as Lord Zuckerman emphasised last year, that human society is itself an important system in the ecological totality. We should pay as much attention to disturbance in socio-economic systems produced by environmental legislation as to changes in other natural systems. It is certainly my belief that we must move forward from the technological base that we have reached rather than to seek a sudden reversal of many of our activities. Thus I am certain that the chemist has a vital part to play in the future technology of the world with its greater emphasis on energy and material conservation and on reduced environmental impact.

Meeting the Needs—the Balance

Annual income twenty pounds, annual expenditure nineteen nineteen six, result happiness. Annual income twenty pounds, annual expenditure twenty pounds nought and six, result misery.

Charles Dickens

Future Economics of Industrial Innovation

by K. Hansen; Bayer AG, Leverkusen, Germany

WHEN, in August, 1973, I accepted your President's invitation to address this meeting, no one anticipated that within a few weeks there would come about in the Near East not only a fresh threat to world peace but also an oil crisis which would have far-reaching and challenging consequences for chemists in particular and, still more, for the chemical industry At a moment when it has just been impressed on us once again how little we know of the future I am here to talk to you about precisely that. And yet it is vital that we should be thinking of the future at all times.

Who at the moment can give reliable answers to the following questions? How scarce will oil be in the years to come? What will it cost? Will we have enough foreign exchange to pay for it? How far will it be possible to use substitutes for oil as a raw material and source of energy? If oil scarcities reduce raw material or energy supplies, where and to what extent should we make cuts or reallocate such supplies as there are? What will be the social effects of such measures? What will be the effects of increases in the price of oil? What classes of goods will be, respectively, the most and least sensitive to increases in the price of oil? How will the utilisation, and hence the profitability, of industrial plant be affected? How will countries react if, owing to increases in the price of oil, they accumulate additional trading and balance of payments deficits? Will they not have to apply a brake by prohibiting imports at a time when other countries are at a loss as what to do with their enormous and still rising balance of payments surpluses? Faced by such imbalances, what will governments do? Introduce still more controls or allow market forces to restore equilibrium? More government interference will not solve the problem. How will zero growth, or even reductions in the output of goods and services, affect the political scene and hence society as a whole? Democracy is all very well if claims to a rising standard of living can be regularly met. If there is no growth, the claims of one group can only be met at the expense of others. Therefore, will inflation break from a trot into a gallop? These are just some of the questions to which there are no definite answers. But all of them need to be answered if predictions concerning the future are not to be mere speculation.

Nevertheless there are unmistakable development tendencies which are occupying the chemical industry, uninfluenced by political and economic crises, and which offer goals whose attainment is worth every conceivable effort—because the future existence of mankind depends on it. It is to these goals, that, with your permission, I shall devote the greater part of my paper. I am referring to the satisfaction of the basic needs of mankind: food, clothing, shelter, and medical care.

Nutrition and health are basic problems of human existence. Regardless of the economic and social system and of the state of development or state of the economy, feeding mankind and the prevention of disease and suffering are elementary needs. At all times these two tasks have spurred the chemical industry on to outstanding achievements. In the future, too, they will continue to be elementary goals necessitating continuous innovation.

Feeding the world's population is no longer an insoluble problem thanks to enormous scientific advances and comparable progress in the skills and technology of distribution. In the last thirty years wheat yields per hectare (ca. 2.5 acres) have increased spectacularly, particularly in the highly industrialised countries. In the USA the yield of wheat per hectare rose from 1.46 to 2.28 tonnes, or 53%; in the German Federal Republic it rose from 2.62 to 4.62 tonnes, which represents an increase of 77%. In the same period the number of persons engaged in agriculture fell by about 80%. As compared with these remarkable yields of agriculture conducted on industrial principles, those achieved in most other parts of the world are very small indeed. The discrepancy shows that the world has sizeable reserves of unexploited food production capacity. Europe and North America together produce more than 40% of the world wheat output on only a quarter of the world's total wheat acreage. In Europe and North America only 98 million people are engaged in agriculture whereas the figure for the world as a whole is 1,900 million. If the present average European wheat yield could be extended to all countries, the world's annual output would rise to 615 million tonnes, which is almost twice as large as the present figure. If the basis for calculation were the average wheat yield per hectare of the United Kingdom, the world as a whole would produce 939 million tonnes of wheat a year, which would represent an almost threefold increase.

There are similar discrepancies in rice production. The average world yield is at present 2.28 tonnes per hectare. The highest national average, 7.38 tonnes per hectare, is that of Australia, where individual yields even reach 10 tonnes per hectare. If the most up-to-date rice production methods, covering everything from the selection and treatment of the seed to fertilisation, pest control, and harvesting techniques, were fully exploited, a yield of 14 tonnes per hectare would be possible. That would enable the world's present rice acreage to yield approximately six times as much rice as it is yielding at present.

Wheat and rice alone now account for 24% of the world's total cultivated surface. In view of the fact that about a third of mankind is short of cereals, it would be possible, if modern agricultural techniques were used consistently throughout the world, not only to meet all demands, but also to achieve a two and a half to threefold increase in the amount of cereals available. It would then be possible to feed 9,000 to 10,000 million people. If the highest yields of the present were achieved universally, the world's total output of cereals would be sufficient for 15,000 million people. Furthermore, if the genetic reserves of nature, as presently understood, were fully exploited in the breeding of new varieties of grain-bearing plants, it would be possible to produce enough cereals to feed a population of 20,000 million. On the other hand, the science of plant breeding is continuing to make such progress that the limits of genetic improvement, far from having been reached already, are so far distant that it is still completely impossible to say where they lie.

The figures I have just given are based on the world's present agricultural acreages and do not depend on such additional environment-modifying measures as the felling of forests, draining of marshes and irrigation of deserts. They could be achieved merely by making full use of the techniques that are already available and are undergoing continuous development: I am thinking, above all, of improvements in soil cultivation, the use of mineral fertilisers and improved varieties of plants, irrigation, pest control, and improvements in the organisation of agriculture.

Potato cultivation trials in which threefold increases have been obtained over the yields obtained by primitive methods lend support to this assumption. Addition of synthetic materials in flock form to the soil can aerate it and reduce its compactness, apart from improving the water economy and utilisation of fertilisers, all of which are conducive to better crops. Trials conducted in deserts and other areas hitherto regarded as uncultivable have produced spectacular results through the use of plastics to suppress evaporation and to make the soil firm and thus resistant to wind erosion. It is therefore not improbable that desert areas, too, will be cultivated in the future.

Highly desirable improvements of the sort I have been considering can be completely reversed by pests. It is therefore necessary to ensure that cultivated plants grow undisturbed by the vast army of weeds, animal pests and parasitic fungi. Pest control is therefore just as important as any other measure taken to improve yields. It begins with the now well established process of seed dressing, continues while the plants are growing and does not even cease when the harvests have been gathered in, for these need to be protected in storage.

The protection and preservation of stocks and the packaging of foodstocks in hygienic plastic materials to enable them to be sold in self-service shops in accordance with changed buying habits are examples of developments which have resulted from changes in social conditions, now that the individual is no longer self-supporting. Consequently, it has been necessary to create efficient distribution systems, particularly in heavily populated regions. Such developments, incidentally, have greatly increased the choice of foods available to us from all parts of the world. They would not have been possible, however, without advances in chemical science.

The shortage of animal proteins is still a problem which is particularly acute in developing countries. Chlorella green algae are very rich in protein, of which, in the dry state, they contain about 50%. Chlorella cultivation trials are therefore being conducted in many countries including the German Federal Republic. It appears that Chlorella cultivation and extraction of the protein is a potential solution to the world's protein deficiency problem. If, at the present stage of development, Chlorella protein is not suitable for human nutrition for purely psychological reasons, it could be fed to cattle and thus be used to increase the supply of protein indirectly.

In 1952 the German, Felix Just, discovered that yeast multiplies in paraffin hydrocarbons and can therefore be cultivated in crude oil. BP in France has developed a method by which yeast can be produced on a large scale and could thus be used to overcome the protein deficiency.

In view of the fact that such ways of raising food output have so far been exploited, either not at all, or to a very small extent, it is reasonable to conclude that the world's food production reserves are by no means exhausted. Indeed, the limits to growth in this respect seem Utopian.

No aspect of life is more closely associated with chemistry than health. The doubling of the world's population which has come about in the last thirty years was caused mainly by enormous increases in the expectation of life resulting from the skill of physicians and scientific advances, not least in the pharmaceutical industry. Even at the turn of the century the average expectation of life was about thirty years in Europe and far lower elsewhere. In the meantime it has more than doubled and in Europe it is now about seventy years. The tuberculosis, puerperal fever, pernicious anaemia, and diabetes mortality rates, to name just a few, have been drastically reduced. In India the number of those stricken with malaria was reduced from 75 million to 5 million in a single decade; the number of deaths per year fell from 750,000 to 1,500. Thanks to improvements in diagnosis and drugs the number of people who reach a ripe old age is much larger than it was a few decades ago; the proportion of pensioners in the population as a whole is increasing steadily. It follows that research and industrial innovation in the chemical industry will have to concentrate on two tasks.

In the first place we shall have to keep up our attack on the remaining scourges of mankind, such as cancer and heart and circulatory diseases.

In the second place we shall see geriatrics becoming a more and more important field of medical and pharmaceutical endeavour. In addition, doctors, sociologists, and psychologists will be called upon to find new and suitable life-styles for old people to ensure that our last years are full of interest and free from degrading poverty.

Pharmaceutical research will be increasingly directed towards these goals. Although the number of drugs available is now very large, there is a genuine lack of drugs for the treatment of many diseases.

The purposes of pharmaceutical research are to discover new active ingredients and to find new indications for those already in use. Its success depends not only on technical skills, good ideas, dedication to science, courage, and endurance, but also, and not least, on luck and the availability of large sums of money. According to the network plans in use at present, the introduction of a new active ingredient involves about 800 separate activities and is therefore a major undertaking, even as far as organisation is concerned. However, large-scale introduction is no guarantee of medical progress, for quality, not quantity, decides the issue.

Chemists, biochemists, pharmacologists, toxicologists, bacteriologists, doctors, and others all have a part to play in the development of a new drug. The first stage is the search for active substances, which have to be synthesised, obtained by synthetic modifications, or isolated from vegetable or animal materials. In the next stage the isolated and purified substances are screened to determine their possible pharmacological effects. By this stage the majority of substances which originally appeared promising have already been eliminated. If a pharmacological effect has been observed, experiments are conducted to find out more exactly what kind of effect it is; tests on various species of animals, and toxicity tests, follow; then, under the most carefully controlled conditions, the toleration of the drug by humans is investigated. After clinical and therapeutical trials the drug is registered, and subsequently released for general use. Assessing the relationship between the beneficial and harmful effects of a new drug is now one of the most important aspects of pharmaceutical research.

Scientific discoveries in every conceivable field and continuous increases in the demands which new drugs are expected to satisfy have made the cost of pharmaceutical research increase from year to year, with the result that it has doubled in the last ten years. On average, the research and development costs of the pharmaceutical industry are now expected to account for 10—15% of sales. On this basis the 1971 figure for the German Federal Republic exceeded 300 million US dollars; in 1970 the figure for the USA, to take another example, was about 600 million US dollars.

Laboratory space for a fully qualified research worker is now estimated to cost about 200,000 US dollars; if the cost of providing the buildings is taken into consideration, this figure rises to no less than 400,000 US dollars. In addition operating costs have to be borne at an annual rate of about 200,000 US dollars per place of work.

Despite this heavy investment the proportion of successes is minimal. In the case of pioneer discoveries, i.e. drugs containing completely new active ingredients, the success quota is 1 in 10,000. Of 10,000 promising compounds, only about 1,500 survive the screening tests and, of these, only about 100 survive biological testing in depth. After the biologically effective substances, numbering about 100, have been thoroughly tested to determine their toxicity only about ten substances remain.

It now takes an average of six to ten years to develop a new drug of high quality and the total cost of doing so may exceed 20 million US dollars. Of this sum, about 30% is consumed in chemical research, about 45% in medical laboratory research, about 10% in clinical development, about 10% in pharmaceutical technology, analysis, and chemical process developments, and about 5% is spent on documentation, literature studies, and general overheads.

Therapeutical progress thus calls for enormous resources of labour and money, while the prospects of success are small and the risks very great.

One observes, looking at the new pharmaceutical ingredients which have appeared in the world as a whole since 1961, that their number has steadily receded. In 1961 eighty-two substances were released. The figure for 1970 was only sixty. In those ten years 755 active ingredients were developed, 24% of them in the USA, 22% in France, 14% in the German Federal Republic, 9% in Japan, 8% in Switzerland and only 6.6% in all other countries together. The Eastern European countries accounted for 2%. In view of this last figure and of the tendency of governments throughout the world to extend their influence over the pharmaceutical industry, even to the point of controlling it completely, it is more than ever appropriate to ask what can be expected of state-controlled research.

The problems I have just outlined are not confined to pharmaceuticals and are found in many other branches of chemical research, such as the development of pesticides. The increasing awareness of pollution that is being shown by governments and the public has already affected the nature of research in numerous sectors of the chemical industry. It therefore seems obvious that chemical innovation, the main ingredients of which are research and development, is going to call for more and more human effort and money and that increasingly large obstacles will have to be surmounted.

In recent years opposition has appeared on the environmental front in particular. But it is far from true that the chemical industry did nothing to combat pollution until forced to do so by the louder and louder cries of its critics. Chemical factories are highly integrated complexes. At the same time chemical manufacturing processes, still more than those of other industries, are sensitive to environmental disturbances. It is therefore a matter of self-interest for chemical companies to operate their plants in such a way that adjacent chemical plants are not adversely affected. Therefore, long before the public became aware of the damage which civilisation was doing to the environment, the chemical industry had devoted a great deal of research and development to the control of pollution; unfortunately the considerable improvements achieved in this way are often deliberately ignored by the public. Environmental research is a very comprehensive undertaking requiring interdisciplinary co-operation between scientists working in an extremely large number of special fields. At the same time it demands exceptional scientific and technical expertise.

As diagnosis always comes before treatment, environmental research has to begin by recognising the problems and recording pollution levels, which means that up-to-date analytical methods and monitoring equipment have to be developed. Wherever possible these should allow analysis and monitoring to be continuous processes. It is interesting to note that sensational reports alleging the poisoning and contamination of foods did not begin to appear until the chemical industry had developed equipment which would enable contaminants to be determined in extremely small concentrations of down to 1 part, or less, in 10 million. Unfortunately the data obtained with these new and sensitive instruments have been interpreted by non-scientists. Thus, as far as chemistry is concerned, there was a boomerang effect, for without advances in analytical techniques it would have been quite impossible to determine such extremely small traces of contaminants.

Equally valuable contributions to the control of pollution are being made in the development of chemical manufacturing processes. A great deal is now being done to modify or even completely change the various stages in industrial manufacturing processes which have been found to cause pollution. Here again the chemical industry can be proud of its achievements. A good example is the double contact process for sulphuric acid production. It enables the sulphur dioxide contents of the waste gas streams emitted by sulphuric acid plants to be reduced to a tenth of their former levels. Because of this advantage the new process has rapidly been adopted throughout the world.

Let me give another example: as you know, solvents play an important part in many chemical processes. But, once used, they are in many cases impossible or very difficult to recover.

Solvents, of course, are an important ingredient of conventional paints and adhesives. However, provided that heat is applied to initiate the chemical reaction, surface coatings can be produced with electrostatic powders and materials can be bonded with adhesives supplied in powder form. In this way solvents are dispensed with.

In the past textile dyeing was always based on the use of water as the cheapest solvent. However, as a proportion of the dye remains in the water instead of being absorbed by the fibres, pollution of river water by effluents from dyeworks has been unavoidable. Thanks to research and development, water has now been replaced in more and more dyeing processes by organic solvents. A relatively low expenditure of energy suffices to evaporate the solvent from the spent dye liquor and return it to the production cycle. So not only is an effluent problem overcome but the dyestuff is more fully utilised.

In the field of pest control, research is tending to supplement purely chemical methods by biological ones. They will involve the use of synthetic forms of the substances with which female insects attract the males. These substances are effective even in exceptionally low concentrations. Destruction, instead of being spread over a wide area, is thus concentrated at chosen locations, which ensures that no harm is done to organisms other than the pest which it is intended to destroy. As there are so many species of insect pests, a correspondingly large number of attractants would be needed for control by this method. This is therefore an enormous field for scientific study.

Benefits to the environment will also result from the highly selective use of pesticides and similar chemicals that are designed to act on specific pests, parasitic fungi, and weeds—possibly at specific stages of their development. Success will depend on a corresponding enlargement of the range of agricultural chemicals available. The problems which must be overcome in their development are similar to those which beset pharmaceutical research.

The chemical industry is spending considerable sums on the prevention of pollution. In the ten years from 1963 to 1972 Bayer alone spent about 1,000 million DM, or about 350 million US dollars, on the operation of plant for the prevention of water and air pollution, as well as spending almost 600 million DM, or about 220 million US dollars, on new pollution control facilities. At the moment about 250 million DM, that is about 90 million US dollars, is being spent annually on new pollution control facilities and the operation of those already in existence. Almost two thirds of the sums spent in this way are being devoted to the treatment and purification of effluents, which is proving to be the most costly pollution control problem. Effluents from chemical factories contain many different substances and these differ in their behaviour on attempted bacterial degradation. We have already isolated strains of bacteria which are very effective in degrading water pollutants— which again shows how much the solution of these important problems depends, not only on the financial resources employed, but above all on the efforts of scientists. This is another case in which future innovations will depend on whether or not financial resources are used in such a way that research and the practical application of its results are correctly synchronised.

Finally, chemical products themselves can contribute to a better environment if, having been discarded, they can be returned to the natural cycle without adverse effects on Nature or on the conditions under which humans live. This again can only be achieved through research and its practical application. Future innovation, as compared with that of the past, will doubtless have to pay more attention to this necessity and methods will be found to eliminate or compensate for other harmful effects of civilisation on the environment. As you may have noticed, I have also dealt with improvements in the quality of the environment under the general heading of the second basic need of mankind, good health.

Additional arable and grazing land on which to grow more cotton and produce more wool to meet the third basic need of mankind is now almost non-existent. The difficulties are increased by the probability that, as living standards rise in consequence of better nutrition, people will want to be better clothed; hence in the course of time the inhabitants of countries which are now still underdeveloped will be more discriminating in their choice of textiles, quite apart from the fact that the total size of the population which has to be clothed will still be increasing. If everyone in the world consumed textiles at the same rate as the average European, today's output of man-made fibres, which meet about 40% of the total demand for textile fibres, would have to be increased roughly six times. Incidentally, as the demand for textiles rises, there is an increasing need for dyestuffs, textile auxiliaries, and detergents. The quality of leather, another clothing material, though one which is becoming increasingly scarce, has been much improved by chemical control of parasites in the appropriate animals. In the long run, however, synthetic leather is the only answer to the scarcity of the natural product. Admittedly, the properties of synthetic leather, such as its ability to 'breathe', are not yet comparable with those of genuine leather; nevertheless, chemists will solve this problem, too, in good time.

Wood is another material which is becoming more and more scarce and expensive—and this brings me to the fourth basic need of mankind: shelter. In housing construction wood is being replaced increasingly by plastics which are designed for their specific application, have better mechanical properties and resistance to weathering, and are easier to maintain. Plastics can now be used for floor covering materials, window frames, blinds and shutters, pipes, bathroom fittings, furniture components, and veneers. Plastics are added to plastering materials to make them last longer. Although the rigid plastics-foam house, produced by a casting operation, is still not yet taken seriously, plastics will be used increasingly in housing construction owing to the scarcity of natural building materials and to the fact that in the long run they are less expensive; and they will result in the development of new building techniques which differ from those for which the traditional materials were suitable. This development can already be seen in industrialised building and the construction of high-rise buildings. The walls of such buildings are no longer built up brick by brick and the walls and roofs no longer consist of the same material. Instead a supporting frame is erected and then, to exclude the weather, the spaces between the girders are filled in with panels which may be made entirely or partly of plastics. Again, synthetic fibres are being used for tent-like roofs for stadia, such as that of the Olympic Games Stadium in Munich, and also for inflated tents.

The examples I have given will, I hope, suffice. I think it should be added, though, that the chemical industry, apart from helping to satisfy the four basic needs of mankind, is contributing a great deal to the high living standards that are now taken for granted wherever they exist and are otherwise so fervently desired. The industry is also playing its part in ensuring that the growth of leisure will be a blessing and not a curse.

And now, I would like to mention some of the factors which, in my view, are indispensable to the preservation of a high standard of civilisation and to the technical and industrial innovation on which this depends. In particular I would mention the scientific knowledge and technical expertise which are among the riches of the human race. Where the individual is concerned, both have to be kept up-to-date through a continuous learning process extending throughout one's life. Scientists and engineers, as one of their obligations towards society, must be prepared to participate fully in this process. To provide adults with opportunities to continue their training, chemical and scientific societies have a duty to organise courses and seminars—as the Gesellschaft Deutscher Chemiker has done on a large scale for many years and with great success.

Even more important is the original training provided by universities and similar institutions for the scientists and engineers of the future, for it is they who will hold the key to scientific and technical innovation in the years to come. Learned societies therefore have the additional responsibility of helping to ensure that courses offered by universities and similar institutions are adapted not only to the demands of the students, wherever these are justified, but also to the future needs of the community. This function must, of course, be a continuous one. It should be added that, where university research is funded by outside bodies, it should be free and uninfluenced by the state or by individuals and organisations who are not primarily interested in extending the frontiers of knowledge.

Until now, training courses at universities have been adapted to progress and the needs of society mainly according to criteria which have been valid in the country concerned. However, to facilitate the comparison of qualifications obtained in different countries, it will be necessary to work towards the standardisation of courses. This is becoming particularly important in Europe because of the high degree of integration already achieved. Our work in this direction will increase the mobility of our professional colleagues. I am therefore glad that the Federation of European Chemical Societies is discussing these matters and that a committee, set up by the Federation at the beginning of 1973, has started an inquiry whose results should form a basis for further progress in this direction.

As scientists and engineers play such an important part in civilisation, they should be particularly aware of their responsibilities and of the social relevance of their activities. In the past there may have been some justification for the pursuit of knowledge for its own sake. But, as full members of society, scientists must now ask themselves whether their work is going to benefit the community. In particular they will have to anticipate and evaluate the possible side effects of their inventions. Will these effects be so harmful that it would be better never to use the invention? Or will the relationship between harmful side effects and benefits to mankind be such that putting the invention into practice can be morally justified?

However, we must also make ourselves better known to the public with a view to convincing society of the importance of scientific and technical endeavour. This again is one of the duties of the scientist towards the community. Among other things, this will entail efforts to ensure that scientists and engineers exert more influence on public life. If they are represented sufficiently strongly in governments and the civil service, they will be able to play an active part in the formation of public opinion and in decision-making processes. Decisions would then be taken according to factual criteria instead of according to purely political or ideological considerations.

In the past scientific and technical innovations have brought great improvements in living conditions, but they have also been used for destructive purposes. It is high time to ensure that they are used exclusively to secure the survival and happiness of the human race. In what I have just said I have therefore tried to make it clear that there must be fundamental rethinking with regard to the pure and applied sciences on which innovation depends and, indeed, with regard to society itself.

Despite the many uncertainties surrounding the economic developments of the next few years—uncertainties characterised by raw material and energy scarcities and by enormous increases in costs—there is no lack of vantage points from which to predict future industrial innovation in the chemical industry, however wide its scope may be. The chemical industry certainly does not lack the courage to meet the challenges facing it. Certain conditions must, however, be fulfilled if we are to be successful. Innovation, no matter in what field, is very, very costly. That is why it is so important that the chemical industry, while mindful of its economic and social obligations towards mankind as a whole, must enjoy the freedom which it needs if its creative forces are to be fully developed, and why fundamental research at universities, which contributes towards innovation indirectly if not directly, should remain free and unhampered. If these demands are met, then pure and applied science, and the industries they serve, will continue to do their part in ensuring that the conditions of life are preserved and improved to the benefit of all mankind.

Education, Training, and Research in Universities

by S. F. Edwards; Chairman, Science Research Council

IN the different countries of the world, Universities have different functions and status. Confining the discussion to the wealthy nations with major industrial backing, Britain comes, in organization, roughly between the United States and Continental Europe, in particular France and Germany. All the advanced countries conduct massive operations in education, training, and research. All conduct their education at the level of first degree for the most part in Universities, even though these have sometimes different names and sometimes, as in the Grandes Ecoles in France, a somewhat different ethos. But their structures in advanced training in science and technology vary more widely, and at the research level there are marked differences.

In the US many colleges do not teach further than first degree and do little research, but the great Universities in addition to all the activities observed in Britain are also often involved in major industrial and defence research, on a much greater scale than in Britain. In Britain, with a few exceptions like the Royal Observatories and the Royal Institution, all pure and basic research lies in the Universities, which also attract quite a substantial amount of support in applied research with contracts from industry and government. In France, Germany, and the USSR, however, most research, even pure research, lies outside the Universities in special Institutes which are funded and staffed independently. All the nations agree that basic research is needed both in pure sciences and in technology, but they organize and pay for it in very different ways. Likewise, these nations give advanced training past the first degree level, but something of the same pattern from the US to USSR is seen. It is also true to say that Britain is exceptionally fast in the speed at which training is completed, and consequently spends far less on it *per capita* than the other countries. Some attempt has been made in the UK to specialise our Universities by having several which started at least as being devoted entirely to science and technology, although the nation drew back from creating the five vast technical Universities originally proposed and settled for a pattern which is closer to the older Universities and which is daily becoming closer still. They all share a great deal of freedom, and not suprisingly again lie in the sequence from the complete freedom (*i.e.* even from direct cash support and advice on numbers) of the US, to the State University of France and the locally controlled Universities of Germany, whose policies are decided politically. Our Universities value their freedom highly and have achieved a remarkable independence especially with regard to finance. Remember that it was not so long ago that the Government Auditor General was first given clearance to review the Universities' books.

Despite their difference in ages and the different ethos in which they were created the British Universities are consistent in maintaining a very high educational standard. They tend to see their *raison d'être* in the words used for the title of my address—Education, Training, and Research—together with a responsibility for scholarship. Indeed, Universities have been rather romantically described as 'Guardians of the Welfare of Knowledge'. By virtue of research and scholarship the corpus of knowledge is increased and by educating and training students the knowledge is disseminated. On the whole Government has been content to allow Universities to manage their own intellectual affairs but has consistently endeavoured to rationalize the quantitative aspect, the numbers of students undergoing further education.

The first review of students numbers was carried out by the Barlow Committee in 1945: 'to consider the policies which should govern the use and development of our scientific man-power and resources during the next ten years and to submit a report on very broad lines at an early date so as to facilitate forward planning in those fields which are dependent upon the use of scientific man-power'. The Committee's recommendation that science graduations be doubled was soon felt to be out of date and Professor Zuckerman, a member of the Barlow team, chaired an enquiry that reported in 1952. Apart from recommendations about numbers the Report contains an interesting point that I will read verbatim.

'There has, however, been remarkable unanimity among our informants that, although present-day science graduates are adequate as scientists, they tend to lack a sufficiently broad education not only in a general sense, but also in the field of science. A scientist whose training has been broadly based can quickly learn other specialities. The narrow specialist is rarely able to step outside the confines of his own particular interests. This is a matter which should be considered further by universities and other training institutions, including technical colleges. It is however, improbable that students at the technical colleges would welcome any broadening of the curriculum they follow, and to which they give up much of their leisure time, unless they feel that such a move is going to be appreciated by their employers as something useful and as something fitting them for advancement.'

I will return to this point a little later. A major contribution was made by the Robbins Committee in 1963 which planned the expansion of the Universities for the 1970s. Recently the Department of Education and Science published figures for student numbers that not only represent an increase on the Robbins estimate (280,000 places by 1976 as opposed to 217,000; 375,000 by 1981 as opposed to 317,000) they represent a new basis for providing higher education.

The Robbins Committee recommended the general principle that courses of higher education should be available for all those qualified by ability or attainment to pursue them, who wished to do so. The new thinking is reflected in the increased number of places which has taken into account what I will call the potential return to the public purse of investment in human capital, and serious manpower planning has now started, aimed at producing realistic figures for student numbers based on the number and type of courses that will be available and the general pattern of employment for the graduates and postgraduates they produce.

This manpower approach is a significant factor which I will return to later. In parallel with the evolution of the Universities there has been a growth in the scientific research they have undertaken. In part this reflects a simple increase in the number of teaching staff able to conduct research, but it also takes into account a change in attitude on the part of the teacher and Government. The adage 'research and teaching go hand in hand' has been lucidly defended by many other commentators and I will do no more than cite it, although I personally have a sympathy for the Vice Chancellor who remarked that they are like beef and milk, related products of the (academic) cow. It is now accepted that at least in the experimental sciences a teacher who is an active researcher is in a position to let one activity bring benefit to the other.

The other point, the growth in funds for research, reflects an enlightened attitude by Government to the rising costs of research brought about by increased complexity of the research itself. I am, of course, addressing my remarks to an audience predominantly comprising chemists; lest any of you feel that my remarks are only valid in the context of accelerator physics or radioastronomy, let me note in passing that to purchase two invaluable tools of your trade, the mass spectrometer and nuclear magnetic resonance spectrometer, will require a sum in the region of £100,000 (should any instrument manufacturer be tempted to offer me a bargain let me specify in advance I require high-resolution double-focusing M.S. and fourier transform electromagnet NMR).

Government expenditure on university research has been both generous and free from strings. The research Councils and their predecessors in DSIR have been able to use part of their funds solely for the support of good ideas—pure curiosity as Lord Rothschild would say—and the White Paper on research has laid down that SRC exists 'to sustain the standards of education and research in Universities'.

We have then a set of self-governing institutions receiving substantial funds from the Exchequer for basic maintenance and for indulging the curiosity of their inmates. Is this a reasonable state of affairs?

It is, I think, fairly clear that while science has been evolving, so too has the public that it serves. Evolving, that is, in the sense of becoming increasingly sophisticated in its demands and more articulate and more successful in putting them forward. Society wants security with personal freedom, employment with leisure, health, communications, transport, education, a clean atmosphere. Society wants all these things at reasonable prices, but it does not mind—or understand—how they are to be achieved. For society, science is a means to an end—a piece of wood is as effective a firelighter as a slow-burning non-toxic high-temperature combustible chemical synthesised by a Nobel Prize-winning chemist.

We have an educational pyramid topped by the Universities within which many of the achievements of chemistry and the rest of science are born, and the demands of society, the members of the public, who have their own criteria for success and failure and their own laws of supply and demand. There is, of course, a third element in the shape of industry—the complex of organisations which is geared to creating and supplying the needs of its markets.

Superimposed on all three elements is Her Majesty's Government. There are Departments for ensuring that society's requirements are considered and ways of achieving them are explored—Health and Society Security, Defence, Agriculture, Fisheries and Food, Energy, Environment, for example. There are Departments charged with overseeing and regulating industry—the main responsibility resting with Trade and Industry. There is the Department of Education and Science with responsibilities for the Universities, and there are sheep-dog departments like the Treasury, the

Civil Service Department, the Cabinet, and Lord Rothschild's Think Tank, who co-ordinate the activities of everybody else.

I do not think it would be appropriate to explore now the interactions between the three elements—interactions which are usually called 'Science Policy'—fascinating as they may be, but I would like to review how the responsible bodies and particularly the SRC has tried to ensure that the needs of Society, the requirements of Industry, and the hopes and abilities of the Universities are brought to bear on each ther.

I would like first to discuss Education and Training. I am not sure whether it is sensible to try to define these topics but my working definitions are that Education is a general intellectual enrichment and development aimed at giving a student better means of thinking; Training is more specific process and is aimed at giving a student tools to apply. During the past 25 years this country has seen a general rise in its overall educational standard and an increasing demand for education. As a result, many professions which used to recruit from schools and carry out in-service teaching now rely on Univer-sities to provide partially-trained manpower (*e.g.* Stockbroking, Banking, and much of Engineering). Many scientifically-based industries and pro-fessions are now running down their non-graduate intake, and a number of new professions are emerging (*e.g.* O.R., Computing Science) that are almost entirely dependent on the Universities for manpower. Although these facts are well known to scientists, it is surprising how little they are appreciated in the country at large. A large part of the public seem to think Universities are exemplified by the study of the classics, and quite apart from the modern developments I find looks of amazement when I explain that the largest group in the 'output' of Cambridge University consists of engineers. The role of Universities in education is markedly different now from what it was 15 years ago. At the same time Universities have come in for a tremendous amount of criticism. Despite the fact that curricula in science have—of necessity—been updated yearly, despite the fact that University research establishes new subjects on the frontiers of knowledge, despite the fact that the products of university teaching find full employment, the Universities are criticised as being out of touch 'with the real world' or 'with the needs of society' or some similar phrase. These accusations are not normally levelled at first-degree graduates as it is usually accepted that the bachelor's degree which normally takes three years in the UK, can only educate students up to a level where they have grasped the fundamentals of a broad field of study and have learned some specialist knowledge.

At first degree level, companies are generally content that they are recruiting above average intelligences without great regard to their need for the specialist knowledge which the recruit brings with him.

There are exceptions such as engineering, economics, statistics, where the graduate is immediately usable. It is at PhD level after a mere 3 years' further work, that the complaints 'too highly specialized' appear. As long ago as 1965 the Royal Society conducted an enquiry into postgraduate training, and the Chemistry Sub-Committee under Professor Raphael examined the specialization question in detail with particular emphasis on differences from and similarities to the American PhD in Chemistry. In the report's own words 'Anyone who expects a cosy panacea from this enquiry, an instant template for guaranteed successful postgraduate training, is bound to be disappointed'. Rather, the report pointed out the danger of specialization turning into parochialization, and emphasized the need to help students appreciate that a PhD is predominantly a training programme at the end of which the acquired skills can be applied in many directions.

SRC conducted a study in 1971 which showed that the production of 'pure research' PhDs in Chemistry was greater that the direct demand, and I have not the slightest doubt that this position obtains for all pure Science subjects. Does it matter? Should we do anything about it? The House of Commons Expenditure Committee is in no doubt that it matters, for in its 3rd Report it suggests that the number of graduate research students should be limited to the number expected to be required to fill teaching and research vacancies in universities, industry, and the public services. I think all the evidence of manpower surveys shows that this would not be a sound basis for building a successful economy. The report also criticizes, *inter alia*, the motivation of PhD students and produces evidence from industry that, if I may paraphrase, 'the wrong sort of PhD is being produced.'

Thus SRC believes that the classical PhD training is a valuable procedure which must be maintained; this is not the same as saying that it is perfect and need not be amended. For the immediate graduate the PhD offers an intellectual broadening and a route to learning how to unearth and evaluate new evidence and principles. He will learn new experimental and/or theoretical techniques. He will take part in formal and informal meetings that will teach him to develop and present his arguments in a rational and coherent fashion. In addition to these points, which I believe are of value and relevance to the Stock Exchange, ICI, or ultimately the Nobel Prize Committee, the student learns the specific trade of surface chemist, natural products chemist, *etc.*, and helps the overall teaching and research effort of his institution. A point completely missed by the Select Committee and by many commentators, is that a very substantial part of the nation's research is carried out by our graduate students, research which in other countries, as I have said, is done at much greater expense in Institutes loosely coupled into the University system.

It is also true that our system is by far the most productive in getting students to an internationally accepted standard in 3 years, so that in weighing the cost of the system its productivity is an essential but often overlooked factor. At the end of his PhD the student is on the labour market and the ease with which he moves into employment varies sharply with discipline. Some of the reasons given by industry (the Expenditure Committee Report contains examples) for not employing PhDs are horrific. I do not belong to the school of thought that says 'Universities know best' but I am struck by the contrast between the attitudes of prosperous foreign companies and their UK counterparts and the difference in composition of Boards of Directors. (How many firms of Stockbrokers would think of hiring scientists to advise on investment?) The Chemical industry, which is one British industry that can stand international comparison, has made its views on PhD training very clear; I wish other sections of the economy would do likewise, for there is often a very large gap between some firms and potential able recruits. I am not yet sure in my own mind whether the student avoids some branches of industry as a result of a mistaken picture of what he will find or *vice versa*; some of each, I suspect. Undoubtedly, many students, having tasted the academic environment, feel that they would like to continue as University teachers, or at second-best as full-time researchers and although some do take up these posts a great number do not.

And yet the country's employment offices are not full of out-of-work PhDs. On the contrary, the PhD in science or technology has no difficulty in gaining employment. Indeed he never did, even in the bleak period of a few years ago, provided he had an open mind on the job on offer. Of course, the occasional protein chemist joins the BBC; of course the occasional gas kineticist joins the Civil Service. But I do not share the view that a PhD not used for research is wasted. These men are utilising the educational component of a PhD. But as we are unable to predict in advance which students are going to take which employment and thereby give them the appropriate vocational training, universities must strike up a relationship with industrial firms to ensure that any gap that does exist between them is bridged more often.

I believe it to be very important for universities to devise schemes of PhD training that will have all the intellectual development of the traditional

PhD, and I believe it to be equally important for industry to establish its manpower needs. If Universities give even more consideration than they are doing to the relevance of their research, many problems will disappear. Within SRC we have tried to encourage this by, for example, our CASE scheme whereby actual research projects are chosen jointly and we have launched experiments jointly with the SSRC in which the whole PhD is oriented to developing the student on a broader front as outlined in the Swann report. However it remains the case that many Departments see themselves first and foremost as generators of research papers and do not stop to consider the education and employment of the student or the obligation they have to the public purse.

So far I have directed most of my remarks at the three-year PhD system; SRC also supports a number of one-year Master's courses which are aimed at giving high-level training in a specific topic. A great deal of effort and care goes into the assessment and selection of these courses on the basis of their absolute quality and the need for their trained manpower. Industrial views go directly to our advisory panels—indeed we have industrial members at all levels in the SRC hierarchy of committees—with the aim of ensuring the relevance of the courses. Courses whose product is not taken up by the industry for which they are designed, are stopped. Referring back to the 'numbers' game an obvious question arises. Undergraduate admissions to science and technology courses are falling below the planned level and the staff numbers are consequently becoming relatively larger. In this country the undergraduate teaching relies to a very great extent on direct teacher-student contact—far more so that is usual abroad—and this is reflected in the staff/student ratio. I do not expect any rapid move to amend the ratio as a result of the decline in numbers, for the education time constant is long, so any move now might not be rectifiable when the trend reverses. But as a result of present circumstances Universities can take the opportunity to begin to run more external courses—short post-experience courses for graduates who wish to learn of new developments in more detail than conferences will allow. This idea was developed in the Docksey Report, and the Chemical Industry, with its own Training Board, is in an ideal position to benefit from the concept. Also, I hope that Universities will take the opportunity to reinforce their PhD courses by formal courses and will encourage interdisciplinary teaching as part of broader PhD training. This sort of teaching is not reflected in crude staff/student ratio statistics.

Let us consider now the role of Universities in research. There is a school of thought that would argue that any such consideration need concern itself with just one topic—leave the universities free to do what they like and discuss how much money they can have to do it. I regret that while this is an attractive argument it cannot expect much sympathy, particularly in the current economic climate. We are all familiar with the arguments used against the practitioners of 'big science' but, as I mentioned earlier, it is merely a question of degree when considering how the arguments apply to chemistry.

You are all more familiar than I with the soaring cost of modern analytical instrumentation; I am sure you are all aware that there will come a point where, just as with accelerators and radio telescopes, SRC and/or the UGC will have to say that a central shared facility is all that can be provided. We have reached and passed this point in some areas—spectroscopy using neutron fluxes or ultraviolet light generated as Synchrotron radiation for example; it is very close in others. Arguments for concentration need not involve just the cost of equipment. Sometimes the expertise required or produced is such that only a handful of installations can be contemplated.

Returning for a moment to manpower problems, there are a number of distinct trends to be seen whereby problems on the boundaries of disciplines are being tackled by mixed teams. For example, chemists understand what reactions occur at surfaces, physicists think they know how to make and study perfect surfaces. As a result joint teams are beginning to pry into the most personal details of a molecule's behaviour at a surface. Similarly, chemists and biologists are probing the nature and mechanism of chemical reactions in the body; chemists, mathematicians, and computer scientists are investigating 'real' as opposed to 'ideal' theoretical problems. Collaborations like these are very warmly welcomed by SRC.

Underlying all research projects, though, is the nature of the problem being studied. In his report Lord Rothschild recognised the need to undertake research with the sole aim of pushing back the frontiers of knowledge. We also warmly encourage University scientists to become involved in the problems arising in industry and in society at large. Now I know that is rather a ritual sentence that everyone says; nevertheless, not much is done other than exhort. For this reason we must warmly welcome ICI's decision to involve a large number of University scientists with their activities. SRC also hopes to encourage more University scientists to work in collaboration with industry, and to give positive encouragement to industrial scientists to spend extended periods in University Departments.

Although I have spoken about Universities so far, at this point, I should mention that SRC has a special concern for the development of the poly-

technics, and has plans to aid the polytechnics to build up a special relationship with the communities they serve. The University system has been
treated generously by successive Governments and is well able, and indeed
eager, to do all the things expected of it. It is therefore pointless to expect
polytechnics to put on PhD courses *etc.* after the University fashion, unless
they wish to be translated into Universities. What they can do is offer
training peculiarly and comprehensively adapted to the area they serve, and
offer the active participation of their staff in the solution of the technical
and scientific problems that arise locally. SRC has published an interim
report on this matter and has set up a committee with the remit to try to find
within three years practical ways in which SRC, in conjunction with the
other Government Departments and Agencies concerned, can help the
Polytechnics to develop new forms of postgraduate education and colla-
borative industrial research.

Finally let me return to the fact that this is the Chemical Society I am
addressing—and that I myself am a Fellow of the Society. Chemistry has
been peculiarly successful in Britain since the war. The Chemical Industry
is one of the most successful in Britain, and its success has not involved large
support from public funds as has, say, the Aircraft Industry. Chemistry in
Universities has maintained a very bright position internationally. SRC has
to help Universities produce the men that industry and education need, and
to help Universities to do the research which will maintain Britain's position.
To do this job we ask many industrialists and government scientists to serve
on our Committees. We value their judgement, but perhaps even more their
ideas. We want everyone to know what we do, and want anyone who has
constructive ideas or advice to feel that he is welcomed in giving it. A major
Government agency like SRC must work in as open a fashion as possible, and
SRC's policy of publishing all its major investigations, all the doings of its
Committees, all the facts and figures concerning its stewardship of men and
money, is one which is functioning properly when the community it serves
takes an interest in what it does and comments constructively. This is particularly the case in so central a subject as chemistry, and I hope as we move into
a tougher period that we will get the support and service from the chemical
community we have had in the past.

Summary

Chairman's Concluding Remarks

We cannot point to a single definitive solution of any one of the problems that confront us—political, economic, social or moral. We are still beginners, and for that reason may hope to improve.

P. B. Medawar

Chairman's Concluding Remarks

by J. W. Barrett; President, The Chemical Society

THE last item in this symposium is 'Chairman's Concluding Remarks'. It has some sort of double meaning if one is not very careful about it! I am not attempting to summarize everything that has been discussed during nearly five days except to comment that I believe the publication of the total symposium that we have had is going to be very useful.

This conference has really been about chemical knowledge—its application to meet society's needs and the implications of the results. Knowledge is an inexhaustible and man-made resource. This is cheering in the light of the problems that we are seeking to solve with regard to energy resources and material resources.

Another point which comes through very clearly both from the programme and the discussions, is that chemical knowledge is totally pervasive. It commands very often the quality of life and meets the material needs of society. From this it must follow that only by its extension and its intelligent use can we hope to cope with the basic needs, let alone some less basic but very desirable needs, of society; and we have been reminded that towards the end of this century the population of the globe is likely to double to the order of 7,000 million.

There has emerged through this conference, not a state of complacency, but a feeling of some degree of confidence that there has been since the beginning of the century a very considerable contribution by chemistry and its related sciences to meeting and satisfying the needs of the world's growing population—moreover that this will continue. Also it is quite clear that the application of chemical knowledge, like any other knowledge, can result in bad effects as well as in major benign effect. One problem is that we have not necessarily been very clever at predicting early enough some of these more undesirable effects in the total living system. Beneficial effect in one part of the system can be paralleled or followed by a disturbance elsewhere.

Reference has been made, of course, and will continue to be made to the example of DDT with its unpredicted chain of events. Here was a new product introduced as a pesticide for the most desirable control of the mosquito. It successfully brought that about, which resulted in a major diminution of malaria in tropical countries. Later it was recognised as being picked up in the living systems of birds and the like. Because of this, although there had been no bad effect on man, DDT was banned in several major countries. Meanwhile the control of malaria it brought, hastened the emergence of free equatorial Africa. No one predicted either the final animal or human effect. Our conference discussions suggest that we should be better able to make predictions of this sort of chain of effects arising from chemical innovation or the application of a bit of chemical knowledge.

Saving 20 million people per year from dying of starvation would seem a very worthwhile objective to anyone, and that has come through loud and clear. It has been pointed out several times that if we could improve world crop yields to those currently being achieved in western countries, then really there would not be any food shortage. The starvation should cease. On the other hand it has also been pointed out that it is not always possible to apply the same remedy to all countries of the world. We need more technology to supplement what is already being effectively used. An example of that is in plant growth promotion, where industry is moving as fast as possible to find out how to help Nature to increase its yields. But it is poorly served with knowledge and theory of the reactions involved in plant growth. We see the urgent requirement of compounding the knowledge and the human assets of university with those of industry.

We have been told that something of the order of 25% of world food is never used because of decay and putrefaction. It has also been pointed out to us that much of this could be prevented by known technology, but it isn't. It has been asserted that the application of certain fungicides is being prevented in countries where there are millions dying of starvation, because of the attempt to apply standards of ideal safety across the world. This is an example of another major concept which has emerged—namely that we are always facing a balance of good and bad, but that the balance can vary greatly from country to country. We can postulate the saving of many from death by starvation by use of a fungicide, but we are prevented from using it because of the belief in sticking to an ideal non-toxic standard for it. The balance has got to be struck and it was notable that this position was put to us by the Director of one of our major research associations concerned with toxicology in the food industry in this country.

We note an intense requirement now for collaboration, both nationally and internationally within industry itself and between industry and government. This was pointed out by Mr. Walters in his very fine paper on energy resources, referring to the fact that there was a welcome and need for the coming United Nations conference on energy resources, that it was no longer sufficient to handle this on a separate-industry or even a separate-nation basis. It was confirmed by Dr. Youle for materials, where the allocation of materials world-wide is something which has now got to be dealt with on an international basis as distinct from only a national basis.

Dr. Holdgate stated the necessity of collaboration on a world basis for pollution control and clean environment, and referred to his very recent attendance at the conference of the United Nations Environmental Programme in Nairobi. However, against all these moves for collaboration arising from the immensity of the problems and the growing understanding of the total needs of society, we are reminded that general solutions are not necessarily right, that there are different situations in the developing world from those in more developed economies, and so one has to be careful not to merge everything in the general international view.

Another important question is the early recognition of possible environmental disturbance. It is a major challenge, a challenge to the chemist, to the biologist, to all those who can harness their present information on a collective basis to give us better predictive capability. Answers can be hastened by the pooling of the talents and capabilities of learned society, government, industry, and university. This was a symposium arranged by the Industrial Division of the Chemical Society. The Chemical Society is a very large society, something of the order of 45,000 members—approximately half of these members being in industry. The Industrial Division is the largest Division of the Society, over 10,000 in membership. It would seem to be clear that the Division is going to be able increasingly to bring together some of these overall talents and capabilities to the advantage of industry and society alike.